T0202076

Sandbows and Black Lights

Sandbows and Black Lights

Reflections on Optics

STEPHEN R. WILK

OXFORD
UNIVERSITY PRESS

Oxford University Press is a department of the University of Oxford. It furthers
the University's objective of excellence in research, scholarship, and education
by publishing worldwide. Oxford is a registered trade mark of Oxford University
Press in the UK and certain other countries.

Published in the United States of America by Oxford University Press
198 Madison Avenue, New York, NY 10016, United States of America.

Library of Congress Cataloging-in-Publication Data
Names: Wilk, Stephen R., author.
Title: Sandbows and black lights : reflections on optics / Stephen R. Wilk.
Description: New York, NY : Oxford University Press, 2021. |
Includes bibliographical references and index.
Identifiers: LCCN 2020040997 (print) | LCCN 2020040998 (ebook) |
ISBN 9780197518571 (hardback) | ISBN 9780197518595 (online) |
ISBN 9780197518588 (updf) | ISBN 9780197518601 (epub)
Subjects: LCSH: Optics.
Classification: LCC QC361.W555 2021 (print) | LCC QC361 (ebook) | DDC 535—dc23
LC record available at https://lccn.loc.gov/2020040997
LC ebook record available at https://lccn.loc.gov/2020040998

DOI: 10.1093/oso/9780197518571.001.0001

1 3 5 7 9 8 6 4 2

Printed by Sheridan Books, Inc., United States of America

All of the following essays originally appeared in different form between 2011 and 2020
in Optics & Photonics News, and are adapted and expanded here with permission
of The Optical Society, with the exception of the following:

Chapter 18: Deck Prisms and Vault Lights
Chapter 40: How the Ray Gun Got Its Zap! II

Contents

Sandbows and Black Lights

Zap! II: Introduction

In the almost twenty years since I began writing essays on strange and quirky optics, I have been through several employers, but in all that time I have stayed a contributing editor for the Optical Society of America. No matter where I was during the day, I always worked on producing these nuggets of infotainment with some regularity. I have always had a backlog of tentative pieces to write, but new topics arose just as rapidly, so I have never been at a loss with a new piece.

The newsletter of MIT's Spectroscopy Lab has, in that time, disappeared, so the essays in this volume are either ones that originally appeared in *Optics & Photonics News* or else have not previously been published in any magazine.

As I stated in the introduction to *How the Ray Gun Got Its Zap!*, my goal was to produce quirky, interesting, and somewhat humorous essays that had a slyly pedagogical edge. "Education by stealth," as the BBC said. In reality, I often start off writing one of these to satisfy myself about some minor mystery of optical science or engineering. I said in *Zap!* that my models were Stephen Jay Gould, Will Ley, L. Sprague de Camp, Isaac Asimov, and James Burke. But I realize now that, in setting the reader a question at the start of the essay—if common wisdom tells us *this*, then why is the given phenomenon doing *this*?—I am following in the lead of two other writers I had encountered. David Webster wrote a book of science and engineering questions, puzzles, and competitions entitled *Brain Boosters* in 1966. It was filled with little experiments and challenges intended to encourage grade-school children to consider things from a different point of view. No math was required, but the problems were challenging enough without it. He wrote a sequel about a decade later. Throughout, he questioned expectations and encouraged what we would now call outside-the-box thinking.

The other writer is Jearl D. Walker. While still a graduate student at the University of Maryland, he put together his book *The Flying Circus of Physics*, in which he challenged the reader to explain various apparently anomalous physical phenomena. In the original edition of his book, he didn't explain these phenomena (although he often taunted the reader with why, based on a simple physical interpretation, the phenomena ought *not* to occur). Instead, at the end of each entry he had a set of reference numbers, which were cross-referenced at the back of the book to a series of books and journal articles. I developed a mental picture of Walker sitting at his desk, going through volume after volume of *The American Journal of Physics* or other relevant magazine. At any rate, I know that I used to do this while supervising my own experiments that required long running periods. They eventually put out a second edition of the book with brief answers at the back, so that students could get a little instant gratification, although Walker encouraged them to seek out the original references for a much fuller understanding.

Walker went on to become a professor at Cleveland State University. He edited the *Amateur Scientist* column in *Scientific American* for a decade, and he appeared numerous times on television. I still have my copy of *Flying Circus* (which has since been republished and has taken up a position on the Internet), occasionally adding notes to new references as I stumble across them.

The point is that both writers encouraged their readers to look at a situation or a phenomenon that appeared to be unusual, or to violate expectations, and to try to explain it or find a solution. I frequently do the same with the topics that are the subjects of the essays in this volume. Why *did* people opt to use a monocle to correct their vision, if it could clearly only fix the vision in one eye? Why *is* the flame of a candle yellow? If you say that it's because of blackbody radiation from the carbon particles in the bright part of the flame (since, as we all know, carbon black has almost perfect emissivity), then you should know that you can't really fit the emission spectrum of a candle to a blackbody curve. If the peak intensity were to fall where the candle's intensity peaks, the flame would be hot enough to melt metal.

What color is brown? If all colors are contained in the chromaticity diagram, then brown ought to be in there somewhere. What determines the values of gray used in a standard illumination chart? They're not steps of equal difference in reflection, and they're not steps of equal difference in optical density—they're equal steps in difference of perceived values of gray. But how do you measure that?

If the public has little interest in the spectrum and spectroscopy (as I wrote about in Chapter 44 of *How the Ray Gun Got Its Zap!*), then how did the notion of susceptibility of vampires to ultraviolet light ever enter the domain of pop culture?

Other articles challenge the reader's engineering sense. Anyone with a little image-forming optics under his or her belt can suggest a way to make a person dressed as a mermaid appear reduced in a fishbowl. But if you're in charge of an attraction like that at a carnival, with no optics training and no access to high-quality spherical mirrors or large lenses, how do you come up with a workable system? More to the point, how do you come up with an optical system that can be built from very inexpensive parts, and which can stand up to repeated disassembly and reassembly as your carnival moves from place to place? Consider it a problem in practical low-budget optics. Similarly, an optics student can probably come up with the dodge that Ted Serios almost certainly used to create his "thought-o-graphs" in the 1960s. These were so convincing that *Life* magazine ran a multiple-page spread on it. But how did Serios come up with it? He wasn't trained in optics, and the solution isn't obvious. What course of thinking could have led him to design his "gizmo"?

Other essays result from forgotten bits of optical history. Jean-Paul Marat wasn't just a French revolutionary—he was a researcher in optics. The widespread use of searchlights as a dramatic, futuristic icon owes much to the head of research at General Electric, who came up with the display as a publicity stunt. The person the Internet says invented the black light (at least as of this writing) did not do so, and never claimed to. The black light dates back to decades before its supposed invention in 1935. Similarly, there was fluorescent paint long before the Day Glo corporation was formed. And 3D movies go back a *lot* farther than most people know—even before 1915, and even before William Friese-Greene's 1890s patent.

Throughout the writing of these, I was helped, as before, by my wife, Jill Silvester, my first reader and critic. I also owe a large debt to the people at *Optics & Photonics News*—editors Stewart Wills and Molly Moser, and creative director Alessia Kirkland.

I thank Susan Denham and Michael Griefe and Laurie Luckreitz regarding William Byler. I'd like to thank Professor Jamie Day of Transylvania University in Kentucky, Deborah Warner of the Smithsonian Institution, and artist Alex McCay for their information on the Claude Lorraine mirror. Brian Rollason and Paul J. O'Pecko gave me useful information on deck prisms. Don Nicholson looked into some questions for "I Was a Teenage Optical Engineer." Tom Kelleher of Old Sturbridge Village and Dick Whitney of the Optical Heritage Museum looked into the background of monocles for me. The late Anthony Siegman suggested the article on acoustic mirrors. Professor Hiromasa Oku of Gunma University not only cited my paper on edible optics, but actually constructed some, and sent me his papers on them.

To these people, and to all the others I am indebted to, as well as the librarians at MIT, Saugus, and elsewhere, my deep thanks.

PART I
HISTORY

1
Who Invented the Black Light?

If you use any search engine to identify the inventor of the black light, and when he did it, you will only find one answer.[1] Dr. William H. Byler invented it in 1935. This factoid is repeated on literally dozens of websites and has even made it into print in a few places. But if you look for a citation, you won't find it. Most places don't even acknowledge that they don't have a cite. Some, like Wikipedia, honestly leave the notation in parentheses—*citation needed*. I could understand some of the websites not knowing where the information came from. It's the normal result of websites copying from each other. But if none of them can pin a fact down to reality, that is a cause for concern.

I had come to ask this question through researching another point that I had hoped would lead to a Light Touch article for *Optics & Photonics News*. The question was: "Why did it take a century from the understanding of the nature of fluorescence until the invention of fluorescent paint?" The query had similarities to the one about the invention of the black light—virtually all websites, and many paper book references, agreed that fluorescent paint was invented by the Switzer bothers in the 1920s–1930s. The Switzers would go on to found the Day-Glo company, which continues to manufacture fluorescent paint. They were honored with a plaque for this work by an engineering society recently. Yet I was able to show that the Switzers were not, in fact, the first to conceive of or to concoct fluorescent paint.[2] Many others had done so, from the late 19th century on. There were even a couple of patents for fluorescent paint before the Switzers did their work. But what the Switzers did was to make a great effort to commercialize fluorescent paints and dyes and to actively search for more of them.

When I started researching the Switzers, I wondered why anyone started this effort to commercialize because an ultraviolet light had just become available. If it was true that the black light was introduced in 1935, then that might explain the Switzers' fluorescent paint. But their efforts began well before 1935, so that couldn't be it.

The fact that the Switzers were developing fluorescent paint before 1935, before the supposed invention of the black light, shows that there must have been easily available commercial sources of ultraviolet light (without visible light) before then. Certainly I knew of laboratory sources—that's how those other people before the Switzers activated their fluorescent paint. It seemed likely that here was another case of the commonly accepted inventor not being the original inventor at all. But if that was the case, then who *did* invent the black light? And how did Byler become the universal choice for inventing it? Did he, like the Switzers, actively promote and commercialize it, this taking a scientific device into the realm of common use?

Most of the Internet references weren't of much help—they didn't tell where the notion had come from. But I began to notice that many of the references seemed to come from the University of Missouri and the University of Central Missouri, and that they seemed to frequently involve the fraternity Sigma Tau Gamma. Byler had been a member of the fraternity, and they listed his invention of the black light as his signal achievement.

Finally I found an article on Byler in the alumni magazine for the University of Central Missouri that referred to Byler as the inventor of the black light, which stated that he obtained a patent on it.[3] Here, finally, was a pin stuck into something solid—a dateable reference asserting Byler's status, and the observation that he had gotten a patent. It would be a simple matter to look through the US Patent Database and find his patent or patents. This would give a fixed date. It would also show how it was constructed. Patents also generally list prior art, so if there had been any previous patents on ultraviolet lights, I would find them listed. Or there might be citations of journal articles.

The Patent Office database is searchable in a number of ways, but the most obvious way is to search for "inventor" and give the name as "William H. Byler." I did this, and found no patents from him in 1935. There were no patents in 1935 for ultraviolet lights, either. Perhaps the year was wrong. Or perhaps he had filed the application in 1935. It often took years for an application to make it through the system and be granted. I broadened the search to any year.

Byler's first patent was dated 1940, and it wasn't for an ultraviolet light. Neither were any of his others. Most of these were granted years later, certainly after black lights were in common use. Looking before 1935, I did find patents for ultraviolet lights, but none of them were granted to Byler.[4] Clearly, he did not invent the ultraviolet light.

Perhaps it wasn't a patent, but a journal article. I used Google Scholar to look for papers by William H. Byler (and W. H. Byler, and William Byler, and other variations). I found several papers, but only one from 1935.[5] None of them were for ultraviolet lights. In fact, two of the papers from the 1940s explicitly stated that in researching fluorescence Byler had made use of an off-the-shelf Sylvania P7 "Blacklite."[6] That, again, seemed odd. If Byler had invented the black light, you'd think that he would be using the one he made. His 1938 paper "Inorganic Phosphors without Metallic Activator"[7] describes excitation with light in the range 330–390 nm, for which a high-pressure mercury lamp with Wood's glass "Blacklight" would have been ideal, but the source he used was a low-pressure mercury capillary lamp using a Corning 586 filter. Again, if he had a black light, why not use it?

I searched through other databases and by arcane methods, and found more papers by Byler, but none on ultraviolet lights. (Google Scholar, though a good and quick and free reference tool, is by no means complete. It certainly doesn't list all of *my* papers. Not by a long shot.) There were a few on fluorescence, and he contributed a chapter to a book on fluorescent paint, but nothing of the sort I was looking for. He was evidently director of the research lab at US Radium in New Jersey for many years, and he was more interested in x-rays and radiation-fueled fluorescence. Not only was there no reference to his working on new ultraviolet sources, he didn't seem to be a particularly likely candidate for inventing such a source. So where did the notion come from?

The article from the magazine had no references, but it did have an author—Michael Griefe. I decided to call the magazine and ask if they could send me Griefe's contact information. It turned out that I was lucky—Griefe was the editor, and he was right there. I spoke with him and learned that he had no recollection of where the information had come from. But, now that I asked, he was curious to find corroboration. He said that he would go to his sources and find which one told of Byler's invention of the black light.

Unfortunately, although his sources—articles from earlier fraternity magazines—do give biographical information about Byler, none of them says a thing about his having invented the black light. At last, Griefe is unable to say where he got the idea.

After looking further myself, I have a theory. Sigma Tau Gamma since 1970 has been giving out distinguished alumni awards. In the year 1988, the award was granted to Byler and specifically listed his accomplishments as "U.S. Radium Corp. Executive Vice President, Leading Luminescent Chemicals Researchers, Inventor of Black Light."[8] One of the fraternity references dates from before Griefe's article, so they can't have copied the idea from Griefe.[9] I suspect that the original award to Byler stated this among his achievements, and people probably seized upon it as the most recognizable thing he had done. Most likely Griefe obtained it from one such listing of alumni achievements, and simply put it into his article, assuming that it must be true, and further assuming that Byler must surely have been granted a patent for important a device. Between the multiple listings on websites of Sigma Tau Gamma chapters, and Griefe's article, the word leaked out into the public sphere, and now his invention of the black light in 1935 has become a meme. People on discussion boards use it as evidence that movies asserting the use of ultraviolet lights before 1935 are clearly in error, because there were no ultraviolet lights then

It still isn't clear exactly how Byler was associated with the invention of the black light. Certainly he did work in fluorescence, and that alone may be enough. It's possible that Sigma Tau Gamma paperwork from the time of Byler's distinguished alumnus award might clear it up, but I haven't been able to obtain that information.

This leaves two questions to be answered: Who *did* invent the fluorescent mercury tube with ultraviolet-transmitting glass surrounding it that has long been the standard "blacklite," used for illuminating blacklite posters, producing fluorescent effects in museums, and other purposes? And what is the biography of William H. Byler, and what did he accomplish?

There are, and have been, many sources of ultraviolet light. Certainly sources of ultraviolet light, separated from visible light, existed in the 19th century. Ritter could not have discovered and demonstrated the existence of "chemical rays"—ultraviolet—in 1801 unless he could isolate it from visible light. He did this using a prism, and using the sun as a source. Later, people produced ultraviolet light using arc lamps, discharge lamps, and burning metal (such as magnesium), and separated the ultraviolet wavelengths using prisms (ideally made of salt or some similar material with low ultraviolet absorption) or diffraction gratings. Clearly, all of this predated Byler, and he had no claim to these.

There are two types of source that throughout most of my life were called "black light," or the hipper "blacklite." These are a fluorescent tube, similar to those used in office lighting, but without the internal phosphor coating that absorbs ultraviolet light and fluoresces in a broad band in the visible. In addition, the glass envelope is made of a glass type that blocks visible wavelengths, only letting the ultraviolet and short-wave visible through. The other type is an incandescent bulb with an envelope made of visible-blocking glass. The latter is far less efficient, since the incandescent filament puts out the bulk of its photons in the visible and infrared, and only a small fraction of them in the ultraviolet. But this type can be use in an ordinary household socket. It seems likely that when it is claimed that Byler invented the black light, the fluorescent type is likely one. I suspect, however, that most people making the claim or repeating it really don't know about the existence of other types of ultraviolet light sources.

Although there were discharge lamps in the 19th century, the beginnings of the modern fluorescent lamp can be traced to the work of Peter Cooper Hewitt, who invented the low-pressure mercury vapor lamp in 1901.[10] This lamp produced a bluish-green light, and it was in many ways clumsier than a modern fluorescent lamp, but it was still more efficient than incandescent lamps of the time, which were still not using tungsten filaments. When they tried to run the Cooper Hewitt lamp at higher currents, it began to get hot enough to soften the glass. They substituted a fused silica quartz envelope for the glass one, but this material let through more ultraviolet light, and people using the lamp began to feel pain in their eyes. The cause was not known, at first.[11] But Professor Schott of Jena began selling such a quartz-jacketed mercury lamp under the name "uviol" in 1905.

Someone who was interested in producing pure ultraviolet light was, ironically, active at about the same time. Robert W. Wood was a professor of physics at Johns Hopkins in Baltimore, and he was an indefatigable investigator of all things optical. He found in 1903 that sending light through a container of nitrosodimethylaniline solution would block the visible components but transmit the ultraviolet. He also found that, optically, cobalt glass combined with a thin sheet of gelatin containing the nitrosodimethylaniline could block most visible light and transmit ultraviolet light, but it wasn't as efficient.[12] At first he used these filters for ultraviolet photography, taking pictures because the human eye wasn't sensitive to those wavelengths. Later he used the filters to filter the visible light out of sources that produced much ultraviolet light, such as arc lamps.

He continued his work, apparently in secret during World War I, coming up with what came to be called "Wood's glass"—a barium-sodium-silicate glass containing 9% nickel oxide. Among other things, a light constructed of a ultraviolet-heavy source screened by Wood's glass would produce ultraviolet light, invisible to the human eye, but which could be detected by selenium photocells or by the fluorescence of substances such as barium platinocyanide. It could be used to transmit secret signals between ships, or between airplanes and airfields, but undetectable to ships not equipped with appropriate detection systems.

After the war, the system continued to be used for signaling but also for demonstrations. Wood used to demonstrate how different things looked under ultraviolet light. One of his favorite demonstrations was to ask audience members to open their mouths. He played the ultraviolet light over their teeth, making them fluoresce. Then he pointed out that caps and dentures were easily detectable, since they did *not* fluoresce, which made everyone instantly shut their mouths.

A similar sort of glass was made by the British optical firm of Chance Brothers, under the name "ultraviolet glass." The early Wood's glass tended to deteriorate over time and exposure to moist air, and it has since been supplanted by other formulations. Corning, Schott, and Kopp today produce similar filter glasses that block visible light while transmitting ultraviolet and infrared light.

Wood's glass got a big boost in 1925 when French dermatologists J. Margot and P. Deveze discovered that they could diagnose several conditions—vitiligo, porphyria, melisma, and others—by observing skin under ultraviolet light. They coupled a light source with Wood's glass, and the result was called a "Wood's lamp," and it is still used under that name to his day.[13]

Combining Wood's glass with a Cooper Hewitt fluorescent lamp was an obvious combination, having a large amount of ultraviolet output.[14]The combination of fluorescent light with ultraviolet filter glass, the combination necessary for a "black light," thus existed by the 1920s. R. W. Wood was the one who appears to have first used a light source with an ultraviolet transmitting/visible blocking filter and even invented a glass for that purpose. If anyone should have the credit for inventing the black light, it's probably him.

Further developments were on the way, however. The Cooper Hewitt lamp used large amounts of mercury, and the light it produced differed significantly from the warm sun-like glow of an incandescent bulb. Work undertaken in Germany and elsewhere in the 1920s resulted in the high-pressure mercury lamp, which used less mercury to produce a brighter glow. And people began using fluorescent compounds to convert the output into something that produced more visible light. French engineer Jacques Risler constructed and patented such a lamp in 1926. Edmund Germer, along with Friedrich Meyer and Hans Spanner, patented a high-pressure mercury lamp with a fluorescent coating in 1927.[15] This led large companies such as General Electric, Sylvania, and Westinghouse to pursue similar devices, improving all parts of the lamp, and particularly finding better fluorescent materials to produce a whiter light. These were successfully made and began to be marketed in the late 1930s. The combination of a new, lower mercury, higher efficiency fluorescent tube without fluorescent material but *with* a Wood's glass-type jacket was the last step required to produce the modern, high-efficiency black light, such as are sold in Spencer's gift stores. I don't know who first put these together, but by 1938 there were experiments done using a high-pressure mercury lamp screened by Wood's glass.[16] By the 1940s, Sylvania was selling them under the name "BlackLight." (The use of the term "blacklight" for "ultraviolet light" long predates the bulb. I have found reference to its use from as early as 1916, and I suspect it's even older.)

William H. Byler was born in Prairie Home, Missouri, on December 16, 1904, one of nine children. He was in the first class to graduate from Prairie Home High School, and he then went on to Central Missouri State Teachers College (now the University of Central Missouri) in Warrensburg, where he obtained a B.A. and A.B. in chemistry and physics in 1927. While there, he pledged Sigma Tau Gamma fraternity and was inducted in 1925. He became a teacher in Ironwood, Michigan, where he met his wife, another teacher. They married in 1929 and returned to Missouri, where Byler enrolled in the University of Missouri graduate school, earning a master's degree in 1931 He then taught chemistry and physics at Hannibal-LaGrange Junior College (Now Hannibal-LaGrange University) in Hannibal, Missouri. He continued his own courses at University of Missouri during the summer and then returned full-time in 1935, having obtained a graduate student assistantship in luminescence research. He obtained his doctorate in 1937, then joined the General Electric Corporation in Schenectady, New York. Two years later he accepted a position as director of research for U.S. Radium Corporation in New Jersey, where he stayed for the rest of his career. He became vice president for Chemical Research and Operations in 1951, and senior vice president in 1967 until his retirement in 1971. He nevertheless stayed on as a director (a position he'd held since 1956) and consultant until 1978, when he moved to Sarasota, Florida. While living there in retirement, he obtained several patents for the relief of conditions associated with aging.

Byler and his wife Thelma had contributed to education at their alma maters. In particular, Byler endowed the Byler Administrative Award at the University of Missouri, which sponsors the Chancellor's Leadership Class, granting scholarship money to incoming freshmen who were leaders in their classes. He also endowed the Byler Distinguished Professor award there, the W. H. Byler Sr. and A.L. Meredith Scholarship Fund, and the Herman Schlundt Distinguished Professorship in Chemistry. At the University of Central Missouri, he endowed the Byler Distinguished Faculty Award, presented annually (it is considered the university's highest honor), and the Byler Faculty Achievement Award. He also endowed a scholarship for students from Prairie Home High School attending University of Central Missouri. He received the Distinguished Service Alumni Award from the University of Missouri in 1972 and the Distinguished Alumni Award from the University of Central Missouri in 1978. In 1983, he received the Sigma Tau Gamma Distinguished Achievement Award, which is the one that mentioned his invention of the black light. He died on December 11, 1985.[17]

Byler worked on the development of new phosphors for radar screens and oscilloscopes, on infrared phosphors that led to the snooperscope, and for x-ray equipment. After World War II, he worked on phosphors for televisions.

As I observed earlier, when Byler had to use ultraviolet light to excite fluorescence, he used a low-pressure capillary mercury lamp with a commercial Corning filter, or else used a complete Sylvania off-the-shelf "Blacklite" source. At no time did he use a source he built himself.

It occurred to me that it *was* possible for Byler to have constructed a high-pressure mercury fluorescent lamp with Wood's glass. He was in a particularly good situation to do so, pursuing a doctorate in physics while on a scholarship to study fluorescent materials, and at the very time that the high-pressure mercury lamp was being developed and pursued for use as a fluorescent light source.

If we accept the year 1935 as canonical, then Byler was still a graduate student at the University of Missouri at the time. He had only submitted one paper for publication, "Use of the Photoelectric Cell in the Study of Phosphorescence," written in collaboration with Albert C. Krueger and submitted to *The Journal of Physical Chemistry* in 1934, but not published until 1935. The phosphorescence studied was excited with a radioactive source, not ultraviolet light, so you would not expect anything about building an ultraviolet source to be there. All his other publications were later.

He was working on his PhD thesis at the time, "Studies on Phosphorescent Zinc Sulfide," which seems to build on his master's thesis, "The Preparation of Phosphorescent Zinc Sulfide," published in 1931. His doctoral thesis was published in 1937, but we would expect it to cover what he was working on in 1935. I haven't obtained copies of these theses, but it appears that the material from both was used to create Byler's 1938 paper "Studies on Phosphorescent Zinc Sulfide 1," published in the *Journal of Chemical Physics*. That "1" seems to be a typo or a mistake of some other sort, for there was no follow-up paper, and it seems to be self-contained. Not all references include it in the title. For a light source for excitation, he used a carbon arc filtered through a Corning 587 glass filter. This, like the one described earlier, peaks at 365 nm. Again, if Byler had assembled a high-pressure mercury lamp with a visible blocking/ultraviolet transmitting filter, this would be the time to use it. That he did not is significant.

In addition, Byler joined General Electric Research Labs in Schenectady just at the time they were feverishly concentrating on producing a high-pressure mercury lamp with an interior visible phosphor coating. It would be a good opportunity to divert material to create a high-pressure mercury fluorescent lamp. But his only paper published while he was at General Electric (aside from the one detailing his thesis work) was "Inorganic Phosphors without Metallic Activator." In this, as noted, he used a capillary mercury lamp and a Corning 586 filter to isolate mostly the 365 nm mercury line, not a "black light."

By 1938, others had produced a high-pressure mercury lamp with Wood's glass, and Sylvania was selling it as the "Blacklite." Byler appears to have had nothing to do with it.

Notes

1. At any rate, at least as of this date I first wrote this—June 2, 2017. Like the original interpretation of the Heisenberg Uncertainty Principle, however, my observation of this phenomenon may perturb it.
2. See Chapter 36 in this volume.
3. The article was "Remembering Blacklight Inventor William Byler" by Mike Greife in *Today Magazine* 9, no. 1 (Summer 2009): 5. The magazine is now called *UCM Magazine*.
4. The patents with "ultraviolet light" in the title were 1,750,024 granted to F. W. Robinson on March 11, 1930; 1,783,643 granted to M. W. Garrett on December 2, 1930; 1,907,294 granted to W. F. Hendry on May 2, 1933; and 1,970,192 granted to H. Lems on February 5, 1935. There were others for lamps emitting ultraviolet light, such as Cooper Hewitt's 1901 patent, that don't explicitly mention ultraviolet light.
5. William H. Byler and Albert C. Kruger, "Use of the Photoelectric Cell in the Study of Phosphorescence," *Journal of Physical Chemistry* 39, no. 5 (1935): 695–700.
6. The articles are W. H. Byler and C. C. Carroll, "A Method for Determining the Chromaticity of Fluorescent Material," *Journal of the Optical Society of America* 35, no. 4 (April 1945): 258–260; and W. H. Byler "Emission Spectra of Some Zinc Sulfide and Zinc-Cadmium Sulfide Phosphors," *Journal of the Optical Society of America* 37, no. 11 (November 1947): 920–922.
7. William H. Byler, "Inorganic Phosphors without Metallic Activator," *Journal of the American Chemical Society* 60, no. 5 (1938): 1247–1252.
8. https://sigmataugamma.dynamic.omegafi.com/wp-content/uploads/sites/535/2014/04/Distinguished-Achievement-Award.pdf
9. That reference is from 2007—http://www.statemaster.com/encyclopedia/Sigma-Tau-Gamma
10. U.S. Patent 682,692, "Method of Manufacturing Electric Lamps," September 17, 1901.

11. W. E. Forsythe, B. T. Barnes, and M. A. Easley, "Characteristics of a New Ultraviolet Lamp," *Journal of the Optical Society of America* 21, no. 1 (January 1931): 30–46.

12. R. W. Wood, "On Screens Transparent Only to Ultra-Violet Light and Their Use in Spectrum Photography," *The Astrophysics Journal* 17 (1903): 133–140. It also appeared in *Philosophical Magazine* 5, Suppl. 6 (1903): 257–263. Wood's paper annoyingly does not mention that the nitrosodimethylaniline is in solution, nor what it is dissolved in. I was initially under the impression that he had somehow cast a screen from this material, which is normally described as a greenish-yellow powder. It is considered a hazardous chemical and a possible carcinogen. It's insoluble in water, but soluble in alcohol and ether.

13. A good reference is Shruti Sharma and Amit Sharma, "Robert Williams Wood: Pioneer of Invisible Light," *Photodermatology, Photoimmunology & Photomedicine* 32 (2016): 60–65. The original dermatologist paper is J. Margot and P. Deveze, "Aspect de quelques dermatoes en Lumiere ultra-paraviolet," *Bull. Soc. Sci. Med. Et Biol. De Montpellier* 6 (1925): 375–376.

14. One such lamp is mentioned as being used in A. C. Roxburgh, "Demonstration of the Detection of Ringworm Hairs on the Scalp by Their Fluorescence under Ultra-violet Light," *Proceedings of the Royal Society of Medicine* 20, no. 8 (1927): 1200; and in A. C. Roxburgh, "The Detection of Ringworm Hairs on the Scalp by Their Fluorescence under Ultra-Violet Light," *British Journal of Dermatology* 39, no. 8–9 (1927): 351–352.

15. US Patent 2,1827,32.

16. *Proceedings of the Rubber Technology Conference, Institute of the Rubber Industry* (1938), 665. https://books.google.com/books?id=j5M5AQAAIAAJ&q=high+pressure+mercury+lamp+woods+glass&dq=high+pressure+mercury+lamp+woods+glass&hl=en&sa=X&ved=0ahUKEwjfgvnk35_UAhUGziYKHWHbBmY4FBDoAQgmMAE

17. Adapted from William P. Bernier (CEO of Sigma Tau Gamma), "Dr. William H. Byler: A Tribute to Our Brother Who Died December 11, 1985," *The Saga of Sigma Tau Gamma* (Winter 1986), 5–6; and from "Two CMSU Graduates Named Recipients of New Distinguished Alumni Awards," *Your University Alumni News—Central Missouri State University* 5, no. 3 (November 1978): 2 and Mike Greife, "Remembering Blacklight Inventor William Byler," *Today* (Alumni magazine of UCM) 9, no. 1 (Summer 2009): 5. https://www.ucmo.edu/today/archives/09/summer/article6_last.cfm. Also material from "Byler Awards Recognize Outstanding Faculty Members," *CMSU News* 2, no. 34 (May 25, 1982): 1.

References

Dissertations

Byler, William Henry. *The Preparation of Phosphorescent Zinc Sulfide.* Master's thesis, University of Missouri—Columbia, 1931.

Byler, William Henry. *Studies on Phosphorescent Zinc Sulfide.* PhD diss., University of Missouri—Columbia, 1937.

Journal Articles

Byler, W. H. "Emission Spectra of Some Zinc Sulfide and Zinc-Cadmium Sulfide Phosphors." *Journal of the Optical Society of America* 37, no. 11 (1949): 920–922. Published while he worked at US Radium.

Byler, W. H. "Methods of Evaluating X-ray Screen Quality and Performance." *Cathode Press* 17, no. 27 (1949): 18–20. Published while he worked at US Radium.

Byler, W. H. "Multibanded Emission Spectra of Zinc-Cadmium Sulfide Phosphors." *Journal of the Optical Society of America* 39, no. 1(1949): 91–92. Published while he worked at US Radium.

Byler, William H. "Inorganic Phosphors without Metallic Activator." *Journal of the American Chemical Society* 60, no. 5 (1938): 1247–1252. Published when he worked at General Electric.

Byler, William H. "Luminescent Pigments, Inorganic." In *Pigment Handbook: Properties and Economics*, edited by Temple C. Paton, 905–923. New York: Wiley and Sons, 1973, 1977, and 1988. Published while he worked at US Radium.

Byler, William H. "Measurement of the Brightness of Luminous Paint with the Blocking-Layer Photo-Cell Used as Photoconductor." *Review of Scientific Instruments* 8, no. 1 (1937): 16–20. Sent in 1936 when he was in the Department of Chemistry at the University of Missouri.

Byler, William H. "Studies on Phosphorescent Zinc Sulfide 1." *Journal of the American Chemical Society* 60, no. 3 (1938): 632–639. Published when he worked at General Electric.

Byler, W. H., and C. C. Carroll. "A Method for Determining the Chromaticity of Fluorescent Material." *Journal of the Optical Society of America* 35 (1945): 259. Published while he worked at US Radium.

Byler, W. H., and F. R. Hays. "Fluorescence Thermography." *Nondestructive Testing* 19 (1961): 177. Published while he worked at US Radium.

Byler, W. H., and H. M. Rozendaal. "The Electrophoretic Mobility of Human Erythrocytes—Whole Cells, Ghosts, and Fragments." *Journal of General Physiology* 22, no. 1 (1938): 1–5. Published when he worked at General Electric.

Byler, William H., and George P. Kirkpatrick. "On the Decay of Phosphorescence and the Mechanism of Luminescence of Zinc Sulfide Phosphors." *Journal of The Electrochemical Society* 95, no. 4 (1949): 194–204. Published while he worked at US Radium.

Byler, William H., and Albert C. Krueger. "Use of the Photoelectric Cell in the Study of Phosphorescence." *The Journal of Physical Chemistry* 39, no. 5 (1935): 695–700. Sent in 1934 when he was at the Department of Chemistry at University of Missouri.

Patents

US 2,266,738. Radio-Active Film. William H. Byler and Clarence W. Wallhausen. December 23, 1941. US Radium.

US 2,454,499. X-ray Screen. William H. Byler and Clayton C. Carroll. November 23, 1948. US Radium.

US 2,487,097. X-ray Screen. William H. Byler. November 8, 1949. US Radium.

US 2,525,860. Phosphors and X-ray Screens Prepared Therefrom. William H. Byler and John W. Wilson. October 17, 1950. US Radium.

US 3,121,232. Color Radiographic Film. William H. Byler, Johanna S. Schwerin, and Frederick R. Hays. February 11, 1964. US Radium.

US 3,224, 978. Tritium-Activated Self-Luminous Compositions. John G. McHutchin, Donald B. Cowan, Ivor W. Allam, and William H. Byler. December 21, 1965. US Radium.

US 3,515,675. Method for Making Luminescent Materials. William H. Byler. June 2, 1970. US Radium.

US 3,631,243. X-ray Film Marking Means Including a Fluorescent Tongue Overlaid with Opaque Indicia. William H. Byler, Halsey L. Raffman, and Frank Masi. December 28, 1971. US Radium.

US 3,717,584. Method for Preparing Rare Earth Oxide Phosphors. William H. Byler and James J. Mattis. February 20, 1973. US Radium.

US 3,845,314. X-Ray Film Identification Means. William H. Byler, Halsey L. Raffman, and Frank Mast. October 29, 1974. US Radium.

US 3,875,449. Coated Phosphors. William H. Byler and James J. Mattis. April 1, 1975. US Radium.

U.S. 3,967,885. Optical Device for Post-operative Cataract Patients. William H. Byler. July 6, 1976. In Retirement, but granted to US Radium.

U.S. 4,012,129. Optical Device for Pre-operative Cataract Patients. William H. Byler. March 15, 1977. In Retirement.

U.S. 4,186,746. Body Warming Device. William H. Byler. February 5, 1980. In Retirement.

U.S.4,214,588. Foot Warming Device. William H. Byler. July 29, 1980. In Retirement.

US 4,249,803. Optical Device for Pre-operative Cataract Patients. William H. Byler. February 10, 1981. In Retirement.

U.S.4,454,869. Arthritis Relief Support Pad. William H. Byler. June 19, 1984. In Retirement.

Final Thoughts

This is as complete a listing of publications as can be made. I checked the publications I found by various means against the Web of Science (formerly the Science Citation Index), and I actually found two more references than they list. I have gone through the papers and the patents and found no references to the innovation of a "black light" source or ultraviolet source in any of them.

2

Revolutionary Optics—Jean-Paul Marat

The Academy has received from a certain Marat
Some theories concerning fire, light, and electricity
This Marat seems entirely certain
That he knows a great deal better than the Academy
....
Light, he proceeds to say, is not light
But a path of vibratorating rays
Left behind by light
Certainly an extraordinary scientist
He goes further
Heat according to him is not of course heat
But simply more vibratoratory rays
Which become heat only
When they collide with a body and set in motionability
Its minuscule molecules.
        —Lavoisier, section 26, lines 138–141, 145–154, in *Marat/Sade* by
                                                        Peter Weiss[1]

I suspect that most people in America who know anything about Jean-Paul Marat mainly have that knowledge from a few bare mentions in their high school or college Western Civilization class, a reading or viewing of Peter Weiss's play *The Persecution and Assassination of Jean-Paul Marat as Performed by the Inmates of the Asylum of Charenton under the Direction of the Marquis de Sade* (usually mercifully abbreviated to *Marat/Sade*, and from which the opening lines of this chapter were drawn), and from Jean-Louis David's painting *The Death of Marat*.[2] He was one of the intellectual leaders of the French Revolution, publisher of *l'Ami du people* ("The Friend of the People"). He suffered from a skin infection that forced him to sit in a medicated bath most of the time. He was visited by Charlotte Corday, who came to visit on the pretext of giving him the names of some wanted Girondists. "Their heads will fall within a fortnight," Marat supposedly remarked. And perhaps they would have, but Corday, a Girondist herself, produced a large kitchen knife and stabbed Marat in the breast, severing his carotid artery and killing him almost instantly. David, a friend and associate, not only arranged the funeral but produced his hagiographic portrait of the slain Marat, which has become an icon (Figure 2.1).

Marat was born in Switzerland. He studied medicine, became a doctor, and developed a following among the aristocracy. In view of his later revolutionary activities, this may seem ironic, but Marat worked through the advantages and restrictions of his

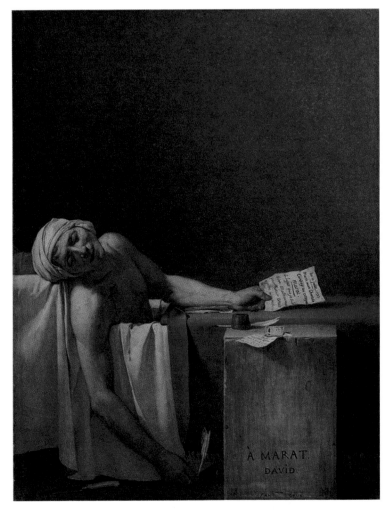

Figure 2.1 *The Death of Jean-Paul Marat* by Jacques-Louis David (1793). The artist was a close friend, who painted this image within months of Marat's assassination. Although painted realistically (and with legitimate details—David had visited Marat the day before his death), this is nonetheless an idealization and a propaganda piece.
Art Resource N.Y.

milieu. Later it was his aristocratic sponsor, the Comte de Maillebois, who arranged for his hearings at the Academy of Sciences. His experimental apparatus was built by the Royal Optician.

In April 1786, he effectively quit medicine and devoted himself to experimental research. He was particularly interested in the hot topics of his day—the nature of fire (the theory of Phlogiston was still a viable one at this time), of optics, and of electricity. To investigate the nature of fire, he employed a device called the solar microscope, which had been in use for some time. It used rays of sunlight, sent in a collimated bundle, to illuminate an object and cast shadows. When Marat used this to look at a candle flame, he was astonished and delighted with what he saw:[3]

> How surprised I was at seeing the image of the candle's flame in the form of a whitish cylinder, bordered by a white halo and crowned with a tuft of swirling jets that were less white.

Marat's drawings illustrating what he saw were published later, and they are almost photographic in quality. It is clear that his solar microscope was acting much like a modern Schlieren system, making variations in refractive index visible. He was seeing what we now know to be the convection of the heated air rising from the candle flame. Marat thought that he was seeing the "igneous fluid," an essential component of both flame and of heat, and that he had made it visible. To confirm that what he saw was not simply due to flame, he immediately substituted a heat piece of iron and saw the same thing. To convince himself that he was, in fact, not seeing some sort of atmospheric effect, Marat observed heated objects within an evacuated bell jar and still saw the rising apparent igneous fluid.

This is one of those issues in the history of science that is rarely covered but really deserves to be. Marat was correctly using the scientific method, eliminating the possibility that he was seeing an atmospheric effect by excluding the atmosphere. The problem was with his apparatus—vacuum pumps at the time were not good enough to achieve really good vacuums. They would not improve for another century, when the diffusion pump was invented. But Marat had no way of knowing this, of course. He requested that the Academy of Sciences review his work. They visited his laboratory and took the measure of his experiments. One of the visitors was the noted American scientist Benjamin Franklin, who volunteered his own baldhead as a subject for the solar microscope. True to form, they saw plumes rising from his head. The Academy issued a very positive report on April 17, 1779. Marat published the report, saying that the Academy supported his work. Unfortunately, this was a misstep—they certainly did not endorse Marat's opinions about igneous fluid—and several members, Lavoisier, Laplace, and Condorcet among them, turned against Marat as a result. The was the beginning of Marat's problems with the Academy.

It didn't help that his next work was a criticism of the work of the highly regarded Isaac Newton. Marat became interested in the phenomenon we today call diffraction. He observed that Newton, while aware of the effect and of Francesco Grimaldi's experiments on the effect, had downplayed the importance of the effect. Marat became convinced that light passing by an edge was somehow being attracted to that edge. He sought to increase the effect by increasing the number of edges, and so he cut multiple slits into cards, passing light through them and onto a screen. These have been called the first manufactured diffraction gratings by some, although they were much coarser than modern diffraction gratings. Marat's efforts preceded those of David Rittenhouse by over five years, but he doesn't seem to have tried to make his slits regularly spaced as Rittenhouse had.[4] Call Marat's constructions *proto*-diffraction gratings.

Among other things, Marat suggested that Newton's interpretation of his own experiment of the breakdown of sunlight into its component colors was incorrect. It wasn't the dispersion of the prism that did so, he claimed, but the effect of diffraction from passing through the narrow slit he used to isolate a band of sunlight.

His work impressed his visitors, and he again asked for a review of his work by the Academy. This time the Academy was more hostile. They took a very long time to issue its report, which consisted of three brief paragraphs. This time they explicitly stated that they did not endorse his conclusions, and that his assertions were "generally contrary to what is best known in the field of optics." It wasn't quite a condemnation. Marat printed the report as a preface to his book on optics and put the best spin he could on it. Wolfgang von Goethe was highly critical of the Academy, accusing it of not being sufficiently critical of Newton.

It's not only with the knowledge of hindsight that we can say that Marat's interpretation had errors—these could have been demonstrated in his own laboratory (in the way, for instance, that slits disperse light on both sides, but the prism only on one side). But Marat had certainly raised interesting points and issues worthy of further examination that the Academy chose not to look into.

Marat himself continued to work in optics, investigating the optics of soap bubbles and translating Newton's *Opticks* into French. But Marat was writing and publishing politics as well, and with the coming of the Revolution in France he started his own newspaper and became a significant and influential voice for the lowest classes and the most radical of policies. Peter Weiss, in *Marat/Sade*, tried to weld Marat's interests together, likening his work with light and heat to his revolutionary activities.

> *He wants to pronounce*
> *the whole of firm and fixed creation invalid*
> *And instead he wants to introduce*
> *a universe of unbridled activation*
> *in which electrified magnetic forces*
> *whizz about and rub against each other.*
> —Lavoisier, section 32, lines 155–160 in Weiss's *Marat/Sade*

Notes

1. The original German title is *Die Verfolgung und Ermordung Jean Paul Marats Dargestellt Durch Die Shauspielgruppe des Hospizes zu Charenton Unter Anleitung des Herrn de Sade* (1964). The version I quote from was translated by Geoffrey Skelton and adapted to verse by Adrian Mitchell (1965).
2. The painting is now at the Royal Museums of the Fine Arts of Belgium. David had visited Marat on the day before his assassination, and so, aside from the idealizations, the picture is unusually accurate.
3. This translation is from Clifford D. Connor, *Jean-Paul Marat: Scientist and Revolutionary* (Humanities Press, 1997; 2nd ed., 2012), upon which I relied heavily for this piece.
4. See my article "Diffraction, the Silk Handkerchief, and a Forgotten Founder," *Optics & Photonics News* 21, no. 10 (October 2010): 16–17; and chapter 9 in my book, *How the Ray Gun Got Its Zap!* (Oxford University Press, 2013).

3
Globulism

Globulism? What's that?

According to some historians of science, it was a mistaken belief in the early 19th century that living matter was composed of spherical "globules" of the same size, acting as building blocks of living matter. But it was all a mistake, due to the aberrations of those early microscopes, and as soon as better microscopes were invented, the higher resolution allowed researchers to see the true and more complex nature of biological tissue. But that's not true, say others. Why would such a theory have taken hold over an entire community? There were underlying philosophical assumptions that made fertile ground for such hypotheses. No, says another—the early researchers were not in error, and this "explanation" in terms of better optical instruments is mere "technological determinism."

So what is the truth? As with everything in the history of science, it's complicated.

To begin with, there isn't really a complete and self-contained movement of "globulism." The term wasn't even applied to this aspect of the history of microscopes until about a century after its supposed heyday. (At the time, "globulism" was a term used in the pseudoscience of homeopathy, with a completely different meaning.) A lot of people looking through microscopes at biological tissues saw globules because, well, here were a lot of globular things to see. Droplets of oil and other immiscible liquids in the water that perhaps suspended the samples. Little bits of matter that had become detached from the main mass. Spherical structures and organs. Cells, possibly. And, undoubtedly, spherical artifacts due to chromatic and spherical aberration. Historians of science are in a difficult position trying to perform a postmortem on centuries-old observations quite simply because there are so many things that the "globules" could be. But let's back up even further.

The microscope was probably invented in the 16th century.[1] Hooke, Leewenhoek, and others reported seeing globules, and undoubtedly they saw a great many spherical items. But no "globulist" theory of biological tissue resulted from their work. Indeed, for more than a century after their time, the use of the microscope seemed to go into a hiatus.

This began to change around the beginning of the 19th century, as several European researchers began to take a deep interest in the structure of tissue. The conquest of chromatic aberration in the 18th century by Chester Moore Hall and John Dollond has been frequently told.[2] Correcting chromatic aberration in microscopes took longer because of the smaller size and tighter radii of curvature of the lenses, but by the beginning of the 19th century there were several achromatic microscopes around. But even chromatically corrected microscopes could—and generally did—have spherical aberration.

George (Jiri) Prochaska (1749–1820) claimed to see globular structure in nerve tissue. Both because there is no such structure in nerve tissue, and because of the description of the appearance that Prochaska gave, medical historian John R. Baker says that what Prochaska was seeing were artifacts due to spherical aberration. In 1823, Henri Milne-Edwards (1800–1885), who would go on to a distinguished career as a zoologist, was a medical student studying animal tissue. He observed globules in many different kinds of human tissue and, remarkably, found them all to be of the same size—1/300 of a millimeter in diameter. To the modern optical scientist, this uniformity of size strongly suggests that the cause was in the apparatus, not in the structures. Over the next three years, according to medical historian J. V. Pickstone, his colleagues Henri Dutrochet and François-Vincent Raspail (who shared his belief in the globular nature of tissues) persuaded him that there were, in fact, variations in the sizes of these globules.[3]

But the idea of uniform globules comprising animal tissues had taken hold, and after 1825 the French biologists Hippolyte Cloquet and Etienne Geoffrey Saint-Hilaire seemed to corroborate Milne-Edwards's observations. The physiologist François Magendie made this uniformity the centerpiece of his theories. On the British side of the Channel, the medic Thomas Southwood Smith and the military surgeon Samuel Brougham agreed. "Every thing susceptible of life may derive all its parts from one constant and primitive molecule, of an uniform character, spherical and colourless," he wrote.[4]

Joseph Jacob Lister was a wine merchant whose hobby was microscopy. His son became the surgeon Sir Joseph Lister, who introduced the use of carbonic acid as a disinfectant in operating rooms, and after whom the mouthwash Listerine™ and the genus of bacteria Listeria is named. To say that the elder Lister had microscopy as a "hobby" doesn't really do him justice—he made significant contributions to the field. In 1823, he observed that the problem of spherical aberration in compound microscopes could be overcome by placing the subject at the aplanatic point of the objective lens and then placing the eyepiece so that its aplanatic point coincided with the second aplanatic point of the objective. In this way all spherical and chromatic aberration could be avoided. Lister had a microscope constructed on this principle in 1823, and he and physician Thomas Hodgkins (after whom Hodgkins disease is named) embarked on a systematic study of tissues.[5] They got into paper skirmishes with defenders of the uniform globule hypothesis.

At first, this appears to be a straight case of superior technology giving greater resolution and revealing the errors of prior work. This was the argument of John Baker.[6] But others, writing about the controversy, were not so sure. Science historian J. V. Pickstone, writing in 1973, commented that "the microscope will not serve to explain the globular theory; artefacts [sic] were not new discoveries, nor was spherical aberration." Pickstone suggests that, while some aberration-dominated observations might have suggested the globular hypothesis, its persistence can only be explained by examining the philosophies of the time, into which such a globular hypothesis fulfilled some expectations.

Another writer, Tom Quick, in his 2011 PhD thesis, sees the British supporters of globule theory, who published in radical journals founded by thinker Jeremy Bentham, as conforming to the ideals of egalitarianism and the democratic ideals of organization.[7]

It seems to me that there might be a simpler explanation. It was only twenty years earlier that John Dalton had proposed his atomic theory, wherein substances were composed of a limited number of atoms, of roughly the same size. What could be more natural than to assume that biological material was composed and organized along the same lines?

Jutta Schickhore, more recently, suggests that the "globular" theorists were not in error at all, and that we are not considering their own work correctly. He acknowledges that Baker shows some fairness to early workers, but is too eager to attribute all problems with the globular theory to the "technological determinism" of the invention of Lister's microscope, and even finds Pickthorn's attribution of the tenacity of globulism to the attractions of different theories to be incorrect.[8]

But Schikhore paints with too broad a brush—Baker always gives good reasons for attributing erroneous observations to spherical or chromatic aberration. And, despite anyone's philosophy, there is the iron fact that one simply cannot meaningfully write or theorize about observations that lie below the resolution of one's instruments. Milne-Edwards and those who followed steadfastly in his 1/300 mm footsteps (even after Milner-Edwards himself retreated from that viewpoint) simply could not say anything about biological structures smaller than that if they could not see them.

Notes

1. Masud Mansuripur, "The Van Leeuwenhoek Microscope," *Optics & Photonics News* 10, no. 10 (October 1999): 39–42.
2. Including in the pages of *Optics & Photonics News*. See Bob Guenther, "Chromatic Aberrations," *Optics & Photonics News* 10, no. 10 (October 1999): 15–18.
3. John G. Pickstone, "Globules and Coagula: Tissue Formation in the Early Nineteenth Century," *Journal of the History of Medicine and Allied Sciences* 28, no. 4 (1973): 336–356.
4. Cited in Tom Quick, "Techniques of Life: Zoology, Psychology and Technical Subjectivity (c.1820–1890)," PhD dissertation, University College London (2011).
5. J. J. Lister, "On Some Properties in Achromatic Object-Glasses Applicable to the Improvement of the Microscope," *Philosophical Transactions of the Royal Society of London* 1230 (1830): 187–200.
6. John R. Baker, "The Cell-Theory: A Restatement, History, and Critique Part I," *Quarterly Journal of Microscopical Science* 3, no. 5 (1948): 103–125.
7. Quick, "Techniques of Life."
8. "Error as Historiographical Challenge: The Infamous Globule Hypothesis," in *Going Amiss in Experimental Research*, ed. Glora Hon, Jutta Schickore, and Friedrich Steinle, 27–45. *Boston Studies in the Philosophy of Science* 267 (2009).

4

Acoustic Mirrors

Along the southern coast of England there are several odd structures, almost all of them inverse hemispheres 10 meters or more in diameter, made of reinforced concrete, most of them slowly crumbling away. They look like giant clumsy television dishes, and there are over a dozen of them. Aside from another in Malta (also built by the British), these are unique devices, constructed between World War I and the early 1930s. They weren't meant as optical devices, but to concentrate sound.

There are many "whispering galleries" in the world. Most of them, like the giant inverted Mapparium Globe in Boston, were not intended for this purpose, and their function is fortuitous. Some ellipsoidal or paired parabolic whispering galleries have been built for museums, such as the Museum of Science and Industry in Chicago, or the Ontario Science Center near Toronto. But the British acoustic mirrors are different from these—they were deliberately built for a practical purpose.

The salient feature of British military history is that it benefits from its isolation from the continent of Europe, with the English Channel serving as a wide and effective moat that severely hampers any invasion. Because of this and a strong commitment to naval power, Britain was able to resist the Spanish Armada in 1588, Napoleon's efforts circa 1805, and the Germans in the two world wars. First Lord of the Admiralty John Jervis, Lord St. Vincent, is supposed to have remarked of Napoleon's troops, "I do not say that the French cannot come; I only say that they cannot come by sea." This implied an image of the French army attempting to invade using Montgolfier-style balloons, and a popular political cartoon of the day shows just that.

A century later, it wasn't so unlikely. In May 1915, German zeppelins and Shutte-Lanz airships crossed the English Channel and dropped bombs on targets on the Humber and the Thames, and later attacked London itself. Within months, German biplanes were crossing as well. During the war they dropped an estimated 300 tons of bombs, producing 5,000 casualties.

Britain couldn't erect an aerial blockade. The solution to dealing with such attacks was anti-aircraft artillery, which had to be brought to bear where and when needed. The earlier the warning could be sounded and the enemy located, the better. The problem is that visual sighting, especially of an intruder whose approximate location is not known, is notoriously difficult. A RAND study from 1965 found that success could be materially improved if the observers knew approximately where to look, but even on a bright and clear day, the probability of sighting fell dramatically after a distance of only five miles. On hazy or foggy days, or at night, such invaders were effectively invisible. Some method of locating enemy aircraft at greater ranges, and in all sighting conditions, was desperately needed.

Radar was years away. Despite the fact that the basic concept had been around for decades, and patents on the idea had been taken out (in Germany, no less), no one really knew or appreciated the capability of using reflected radio waves for ranging. The first observation of moving aircraft with what was essentially a radar system did not take place until 1930, and then it was a fortuitous accident.

Within two months of the German air attacks, the British were pursuing a solution, under the direction of a Professor Mather. They cut a semispherical depression into a horizontal chalk wall at Binbury Manor. A listening trumpet was placed on an adjustable mount near the focus, and an operator listened to the concentrated sound using rubber tubes connected to the trumpet, stethoscope style. The operator moved the trumpet around to maximize the sound, and the azimuth and elevation could be read off grid marks. This explains why the depression wasn't parabolized—the mirror itself wasn't movable, so "aiming" had to be done by moving the detector around, and it was best to keep the depression symmetric.

Professor Mather claimed that this device would detect audible sounds from as far away as 20 miles, and aircraft engines were certainly noisy enough. The Army did their own tests, disagreed, and threatened to cancel the program. (Mather thought the testers were incompetent.)

In any event, the program was continued, possibly because no one had anything else to suggest. More mirrors were built, larger ones, and the surfaces were lined with concrete for better sound reflection. Later, the mirrors were built free-standing entirely out of reinforced concrete. In 1917 and 1918, the mirrors near Dover proved their worth, detecting airplanes at a distance of 12 to 15 miles. In October 1917, airplanes headed for London were detected in time to give several minutes' warning.

After the war ended, the experiments continued. There was still no other solution to the problem of long-range aircraft detection. Larger acoustic mirrors were constructed, and microphones were tried, although stethoscopes were preferred. In 1925, Dr. W. S. Tucjer was put in charge of the program, and he started building 20-foot mirrors and then 30-foot ones along the south coast, along with structures for personnel and equipment. Booths were built for operators, who controlled the detector position with hand wheels and foot pedals. Eventually mirrors as long as 200 feet were built to detect long wavelengths (although these were not complete hemispheres). With practice, they could detect aircraft at a range of 30 miles.

The operation was not without considerable difficulties. The concentration required of the operators meant that shifts could be no longer than 40 minutes. Wind blowing across the mirror could mask faint sounds, and curtains were hung across the ends to decrease the effect. Ambient noises in front of the acoustic mirrors could be picked up, interfering with aircraft detection (the mirrors were not all built at the edge of land, looking out over the water), although the story that a passing milk truck ruined an inspection of a listening facility is apparently apocryphal.

Plans to build an extensive series of mirrors were proposed in 1935, but they were derailed by the coming of radar. In July 1935, a radar installation that had been set up at Orford Ness was detecting aircraft at a range of 40 miles, and without strain on the operators. The construction of further acoustic mirrors was cancelled, although operation of the existing ones continued. The last ones were not phased out until 1939.

Even so, their career was not over. Into World War II their operation provided a useful "cover" for British use of the still-secret radar systems. When German scientists began jamming radar signals later in the war, it was thought that the acoustic stations might be used until a solution to the jamming could be found.

Today, most of the acoustic mirrors still exist. Some are buried. A few have been destroyed by time and weathering. Many more are on private land, and not legally accessible. Many of the ones that are reachable have been covered with graffiti. Efforts are under way to preserve them. They have been used in videos and works of conceptual art.

In fact, modern artists have started building new acoustic mirrors in public parks. To my knowledge, these only exist in Britain, the only country to actually use the technology on a practical basis. Paired listening and speaking concrete mirrors have been built along the Palmarsh Footbridge over the Royal Military Canal in Kent, not far from a set of the original World War I mirrors. Another set, also of concrete, was erected at Wat Tyler Country Park in Essex, not far from the coast. Yet another set, this pair of aluminum, and with a parabolic profile, is in River Colme Sculpture Park. Finally, more concrete sound mirrors are at St. James's Mount in Liverpool.

References

"Acoustic Mirror." *Wikipedia*. Accessed April 27, 2011. http://en.wikipedia.org/wiki/Acoustic_ mirror

Dyckhoff, Tom. "I'm on the Beach." *The Guardian*, June 13, 2001. Accessed April 28, 2011. http://www.guardian.co.uk/culture/2001/jun/13/artsfeatures.arts1.

Gillie, Oliver. "Listing of Sound Mirrors Urged." *The Independent*, July 3, 1993. Accessed April 28, 2011. http://www.independent.co.uk/news/uk/listing-of-sound-mirrors-urged-ollie-gillie-reports-on-the-preradar-detection-devices-that-enthusiasts-want-to-see-preserved-1482621.html

Grantham, Andrew. "Royal Military Canal Sound Mirror." http://www.andrewgrantham. co.uk/soundmirrors/other/royal-military-canal/

Grantham, Andrew. "Wat Tyler's Sonic Marshmallows." April 6, 2008. http:// andrewgrantham.co.uk/soundmrrors/tag/sound-mirror-art/page/2/

Hyde, Phil. "Sound Mirrors on the South Coast." January 2002. Accessed April 27, 2011. http://www.doramusic.com/soundmirrors.htm

"A Sound Installation by Matthew Sansom." http://www.matthewsansom.info/colnevalley. htm

5
Friedrich Richard Ulbricht's *Kugelphotometer*

William Manchester's history of the ending of Middle Ages, *A World Lit Only by Fire*, covers the historical period of about 500 CE to 1521, but it could just as easily refer to civilization until the end of the 18th century. Aside from intellectual curiosities like bioluminescence and triboluminescence, humankind depended upon fires, candles, and simple lamps until around the turn of the 19th century. That century saw an accelerating and dramatic change in the way people lit their homes, their businesses, and their streets.

It all started changing when the Swiss physicist and chemist François Pierre Ami Argand invented and patented his new design for a lamp. The important part of the design was the hollow cylindrical form of the wick that allowed passage of air through the center, and a glass chimney to direct the flow of air. Together with a mechanism for raising and lowering the wick, this constituted a revolution in lighting. The Argand lamp burned more brightly than an ordinary lamp or a candle, producing the equivalent light of many candles—five or six or seven or eight candles (depending upon whom you talked to). Furthermore, it burned the fuel completely and cleanly, and it was less expensive than an ordinary lamp. With modifications, it could burn almost any type of oil, from whale oil to Colza oil (a plant extract) to vegetable oil to paraffin to kerosene. The Carcel burner, a modification of the Argand lamp that used a clockwork pump, allowed it to have a non-gravity-feed reservoir. The Argand lamp was a revolution in lighting and an underappreciated technological advance. British inventor Sir Goldsworthy Gurney added a jet of oxygen to the central hollow and produced a much more intense white flame. His Bude burner was used to light the House of Commons for over half a century.

The nineteenth century also saw the invention of limelight, also by Gurney, in the 1820s. The limelight or Drummond light used an oxyhydrogen flame playing upon a ball or cylinder of calcium oxide (quicklime), which glowed with a bright and very white light due to the principle of candoluminescence. Its use was restricted to theatrical lighting, but it was a major leap in the production of bright light.

Of course, the most profound effect was the development of electric lights, first the arc light of Humphrey Davy at the beginning of the century (although it would not see commercial use until the 1880s) and then the incandescent lamp, worked on by many and made a commercially practical device by Sir Joseph Swan in England and by Thomas Edison in America, both in 1879.

A very important, if less impressive, technical achievement was the spread of gas lighting, used to light public streets and private homes, using coal gas produced at a central source and distributed through pipelines. Gas lights started in the first decade of the 19th century and spread through European cities in the 1820s.

Less appreciated, but still important, was the invention of the thorium gas mantle by the Austrian chemist Carl Auer von Weisbach in 1885–1891, which multiplied the visible light output of any flame. Coupled with an Argand lamp or a gas lamp, this provided a quantum leap in illumination to areas electricity still did not serve.

Mostly forgotten today are the Nernst lamps invented by German chemist Walther Nerst in 1897. Using ceramic filaments instead of metal, these lamps provided a more natural light and didn't require vacuum or noble gas coverings. There had also been experiments with fluorescent lighting, although that would not become commercially feasible until the 20th century.

Another undeservedly obscure development is the creation of the modern candle in the 19th century. Michel Eugène Chevreul and Joseph-Louis Gay-Lussac isolated stearin from animal fats and patented it in 1825. It burned more cleanly than traditional animal fat candles and could be added to other candle bases to produce a harder candle. Paraffin candles were first introduced in 1848, providing inexpensive candles with low odor and residue from nonanimal sources.

This number and variety of lighting sources inevitably bring with it questions— Which of these sources is best for an application? Which is brightest? Which is the most efficient? It dates back to the Argand lamp, which clearly produces more light than a single candle or lamp flame—but *how much* more? As noted earlier, estimates varied. More to the point, how could you measure or objectively verify this?

There was much interest in measuring light in the late 18th and early 19th century, with photometers invented by many researchers, including Count Rumford, John Herschel, Johann Heinrich Lambert, and Robert Bunsen, whose photometer was probably the most widely used. All of these devices, however, relied ultimately upon the intensity-resolving power of the human eye. Until the invention of photography, there was no other method of measuring light intensity. And photography was a slow and clumsy method, which ultimately relied on the human eye to tell which of two recordings was darker. Variable star observers relied upon human eye judgments of relative intensity (and still do, in amateur circles), but these rely on observations of relatively dim sources. When judging relatively bright sources, practitioners realized that the human eye is easily dazzled, undergoes fatigue, and is generally unreliable.

The discovery in 1873 by British electrical engineer Willoughby Smith that the conductivity of selenium changed proportionally with the amount of light it was exposed to led immediately to the development of the electric light meter. There were still issues with consistency and drift, but the measurement of light had been freed from its dependence on the human eye and the slow process of photography. Now the measurement of the relative output of different sources of light could be undertaken. Or could it?

Putting a standard detector in front of a source, even at an agreed-upon distance, doesn't give a satisfactory answer. Many sources vary with angle and azimuth. And, as some engineers pointed out, what's really important is how much light is sent into the environment. Devices were built using sets of mirrors or ellipsoidal mirrors to direct all the output from a source onto a single surface, and the light reflected from that measured. It was difficult to satisfactorily gather this light using such arrangements. And then, two workers independently showed that it wasn't really necessary.

W. E. Sumpner was an electrical engineer and physicist, the "principal" of the Birmingham Technical College in England for 35 years. In 1892, he suggested that it ought to be possible to measure the flux of light by letting it undergo multiple diffuse reflections within a chamber with low loss but matte reflective surfaces. The multiple reflections would cause the light to "even out" and lose its directionality. Introducing

the light through a small aperture and using a detector elsewhere, after multiple reflections, should give a result correlated with the input flux. No exotic collection
method was needed to gather all the light onto one surface. Sumpner did not perform
the experiment, however.

Friedrich Richard Ulbricht was a professor of electrical engineering at the
Königlich Sächsischen Technischen Hochschule (today the Technische Universität)
in Dresden, Germany. In the early 1890s, he wanted to determine the most efficient
method of illuminating Dresden's train stations using electric lighting. He was faced
with the problem of measuring and comparing the different candidate systems. His
response to the technical challenge was the same as Sumpner's, but, as he had a practical problem to address, he put it into action. He constructed a hollow sphere half a
meter in diameter and coated the interior with highly reflective white paint.[1] He called
it a "globe photometer" (*Kugelphotometer*). Today it is called an "Ulbricht sphere"
(*Ulbricht-Kugel*) in Germany and an "integrating sphere" in English-speaking countries, and it has become the standard method of measuring absolute flux levels and
reflectance, and for providing the best real approximation to a Lambertian source.

Note

1. Early spheres used magnesium oxide, which Ulbricht probably used. Later barium sul-
   fate was used, and later still polytrifluoroethylene (PTFE), variations of which are still
   used today.

References

References on the History of Photometry

DiLaura, David L. "Light's Measure: A History of Industrial Photometry to 1909." *Leukos*
1, no. 3 (January 2005): 75–149.

Johnston, Sean F. *A History of Light and Colour Measurement: Science in the Shadows.* Boca
Raton, FL: Institute of Physics Publishing, 2002 and 2015.

Johnston, Sean F. "Making Light Work: Practices and Practitioners of Photometry." *History
of Science* 34 (1996): 273–302.

Johnston, Sean François. "A Notion of Measure: The Quantification of Light to 1939."
PhD Thesis, University of Leeds, Department of Philosophy, Division of History and
Philosophy of Science, November 1994.

Otter, Chris. *The Victorian Eye: A Political History of Light and Vision in Britain, 1800 –
1910.* Chicago: University of Chicago Press, 2008.

Ulbricht, Friedrich Richard. *Das Kugelphotometer.* Berlin: R. Oldenberg, 1920.

Sumpner's Papers

Sumpner, W. E. "The Diffusion of Light." *The London, Edinburgh, and Dublin Philosophical
Magazine and Journal of Science* 35, no. 213 (1893): 81–97.

Sumpner, W. E. "The Diffusion of Light." *Proceedings of the Physical Society of London* 12, no. 1 (1892): 10–29.

Papers on the Theory of the Integrating Sphere (the following list is by no means complete, as there are numerous papers)

Benford, Frank. "The Integrating Factor of the Photometric Sphere." *Journal of the Optical Society of America* 25, no. 10 (October 1935): 332–339.

Cowther, Blake G. "Computer Modeling of Integrating Spheres." *Applied Optics* 35, no. 30 (October 20, 1996): 5880–5886.

Goebel, David A. "Generalized Integrating-Sphere Theory." *Applied Optics* 6, no. 1 (January 1967): 125–128.

Hardy, Arthur C., and O. W. Pineo. "The Errors Due to the Finite Size of Holes and Sample in Integrating Spheres." *Journal of the Optical Society of America* 21, no. 8 (August 1931): 502–506.

Jaquez, John A., and Hans F. Kuppenheim. "Theory of the Integrating Sphere." *Journal of the Optical Society of America* 45, no. 6 (June 1955): 460–470.

Prokhorov, Alexander V., Sergey N. Mekhontsev, and Leonard M. Hanssen. "Monte Carlo Modeling of an Integrating Sphere Reflectometer." *Applied Optics* 42, no. 19 (July 1, 2003): 3832–3842.

Ross, E. B., and A. H. Taylor. *Theory, Construction, and Use of the Photometric Integrating Sphere.* Washington, DC: US Government Printing Office, 1922. https://nvlpubs.nist.gov/nistpubs/ScientificPapers/nbsscientificpaper447vol18p281_A2b.pdf

Whitehead, Lorne A., and Michele A. Mossman. "Jack O'Lanterns and Integrating Spheres: Halloween Physics." *American Journal of Physics* 74, no. 6 (June 2006): 537–541.

6

The Monocle

The monocle seems more like a stage prop rather than a serious optical device. It was the not-quite-serious indicator for the upper crust and elegance. British aristocrats wore monocles and so did Charlie McCarthy. In advertising, putting a monocle in the eye of your mascot suggested quality. Mr. Peanut wore one. So did the tomato-headed character promoting Heinz Tomato Juice in the 1930s and 1940s. But that symbol of aristocracy was, for many comedies, mainly a way to puncture their pomposity by having it drop unexpectedly from the orbit to swing at the end of its lanyard, or to fall into the soup.

In the 1930s and onward, the monocle also acquired an association with villains—Erich von Stroheim wore one in several movies, as did Batman's enemy the Penguin, or Rocky and Bullwinkle's nemesis Fearless Leader. It was associated with German generals in World Wars I and II, including Erich Ludendorff, Werner von Frisch, and Hugo Sperle. Such association with the military and staff of a war foe soured many people on the monocle, and it disappeared for many years in America.

Undoubtedly it was often worn as a fashion accoutrement, a symbol of elegance and style, and possibly with an erotic charge. No one believes that Madonna needed optical correction when she wore a monocle onstage, or when Lady Gaga copied her. There was even a 1915 article entitled "The Monocle as Erotic Fetish."[1]

Most optical shops do not sell monocles. Clearly the lens can only have an optically corrective action on one eye, while most people generally needing optical correction require it in both eyes. Add to this that a great many of the monocles sold have zero optical power, and one has to ask if there is any serious ophthalmological point to the monocle at all.

It turns out that there is, but it's not an obvious one, and to understand the development of the monocle as a corrective vision device, we need to understand fashion, technology, and history together. There were single lenses used for optical correction when lenses began to be used for vision correction, but that was likely because lenses were expensive and difficult to make, especially negative lenses for correcting myopia. Pairs of spectacles became common at the end of the 17th century, and monocles before this time were probably a matter of affordability and availability.

Once pairs of spectacles became available, they became items of fashion as well as practicality, and their wearing had implications. Gentlemen did not like to wear spectacles. It was a sign of weakness. When George Washington addressed his potentially mutinous officers in Newburgh, New York, on March 15, 1783, he appealed to their memories of the Revolution and their sense of duty by reading to them a letter while wearing his new and rarely worn spectacles, saying "Gentlemen, you will permit me to put on my spectacles, for I have not only grown gray but almost blind in the service of my country."[2] Jonathan Swift's Lemuel Gulliver had spectacles in Lilliput, but he did not wear them until he needed to protect his eyes. The tradition continued on into the

20th century, with famous and powerful men (such as John F. Kennedy and Dwight Eisenhower) not wanting to be photographed in eyeglasses.

Under these circumstances, it could be acceptable to use a single corrective lens, especially if it was not used in a permanent setting affixed to the bridge of the nose, but was a temporary thing, held in the orbit of one's eye by pressure between the cheekbone and brow. Such an item might be an affectation and dismissed as such. If it was a positive lens, one might use it as a portable magnifier.

In fact, at the end of the 18th century and the beginning of the 19th century, that is precisely what people did. They carried small lenses (not only round ones but also rectangular and octagonal ones) with short handles. These were often heavily decorated, and often bejeweled, and were called "quizzing glasses." Here "quizzing" meant "looking" or "examining." These were associated with aristocratic dandies and were often the object of fun. Quizzing glasses began to be replaced by monocles after about 1830. Even though the monocle was supposed to be worn screwed into the eye, it was also used as a magnifying glass. Think of the mascot of the *New Yorker*, Eustace Tilley, in his characteristic pose of examining a butterfly through his monocle.

Pitt H. Herbert, a lens designer for half a century and collector of antique spectacles, presented an interesting look at the history of the monocle.[3] He claimed that monocles first made inroads among people in the theater, and from there spread to the upper classes, mostly as a matter of style. His article claims that after 1820 such eyewear began to be manufactured in more styles and a range of prices, enabling anyone who wanted a monocle to purchase and wear it. To facilitate the wearing, the gallery was invented—an extension above the top of the monocle and perpendicular to the surface, it allowed the brow to grip the monocle, yet keep the surface away from the eye. The monocle craze lasted only about three years, but monocles continued to be worn throughout the 19th century and into the 20th. During World War I, the British government included monocles on the list of Essential Production Items (Figure 6.1).

The monocle was introduced to America, Herbert says, by about 1880, and acquired the nicknames of "dudes" and "oxfords." The fad reached its height in the United States about 1913, and monocles were produced with rims in such unlikely materials as celluloid and hard rubber.

How can the monocle be a serious eye corrector if it only corrects one eye? Another question is why the statistics of monocle use don't match the needed eye correction of the users. Depending on which study you go by, there are slightly more myopic people than hyperoptic ones, although presbyopia tends to make older people slide toward farsightedness. But an informal survey of monocles on the Internet shows that, although negative power ones exist, positive powers are far more common. In fact, NearSights, the Internet seller of monocles, only has its online eye test work for positive powers. The same holds true for the quizzing glasses I've found. What is a myope to do? Pope Leo X had his myopia corrected with a negative monocle, but the modern American may have to hunt around to find one.

Doesn't it do harm to the eyes to only correct one of them? That was the feeling of doctors in the first part of the 19th century, who wrote articles against the use of the monocle, warning that this pernicious habit would lead to an "unbalancing" of the eyes. It would seem to make sense to use a monocle if you only had one eye, but the only single-eyed wearer of a monocle I know of is Inspector Kemp from the movie

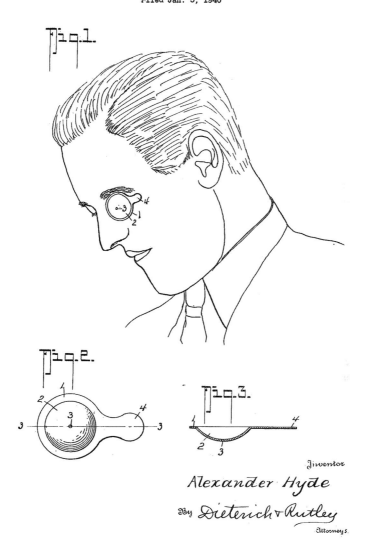

Fig.1

Fig.2.

Fig.3.

Inventor

Alexander Hyde

By Dieterich & Rutley

Attorneys.

Figure 6.1 U.S. Patent Office drawing of a pinhole monocle, 1941.

US Patent 2,230,185. Courtesy of the United States Patent Office

Young Frankenstein, and he wore his monocle over the eye with the eyepatch, for comic effect.

But there is an unexpected truth buried in all of this that ties it all together and makes the monocle make sense—the monovision effect.[4] There are references to this as early as 1942, when the *Optician's Yearbook* suggested it as a desirable optical practice, but the very success of quizzing glasses and monocles suggests that people had already discovered it much earlier. If one uses correction on the dominant eye alone, while viewing with both eyes, the brain will coalesce the images in such a way that the clear image will be evident as if seen by both. (In fact, it was found that one could correct each eye to a different prescription—using one for near vision and the other for far vision, and the brain can select the view desired.)

Rather than unbalancing the eyes, then, the use of single-eye correction produces effectively normal vision, without the encumbrance or social stigma of donning eyewear for both eyes. Furthermore, the use of a positive lens corrects the farsightedness that increasing age brings on, so it's ideal for older individuals of high status who want to be able to see as they used to. It's thus not too surprising that this became the eyewear associated with those wealthy aristocrats. The "dandy" implications, I think, kept American politicians away from this, but British politicians wore them, as did noted actors and artists like Toulouse Latrec, Fritz Lang, and others.

Today, you can get the advantages of wearing a monocle without its being obvious. Eye surgeons will sometimes use different intraocular lenses in each eye of a patient, so that they may have effective bifocal vision by doing the sort of brain-switching mentioned earlier. The same thing can be done by using a single contact lens on one eye, which led one doctor to call this "the invisible monocle." One recent article even claims to hail "the return of the monocle."[5]

Regardless of the medical benefits, the monocle is restaging a low-level fashion comeback as well. With buying easy over the Internet, one can find monocles more easily now than in the past. Furthermore, the "steampunk" style revels in things seen as 19th-century retro, and the monocle is, for most young practitioners today, a thing of the past that fits perfectly into that category. I see numerous sites selling monocles for the steampunk crowd. Many of these are recycled Diopter sets from companies like American Optical—I can not only identify the size and style, but the handles, with their cutout "+" or "−" signs (to distinguish positive from negative lenses) are clearly visible, even though they may have restraining ribbons threaded through them.

Notes

1. R. K. Neumann, *Zeilschriflf. Sexualmisscnschaff* (January 1915), reported on by James O. Kiernan, *The Urologic and Cutaneous Review* 19, no. 5 (May, 1915): 287. http://books.google.com/books?id=qyE2AQAAMAAJ&pg=PA287&dq=monocle+erotic&hl=en&sa=X&ei=eYVtUtbFJ8rwkQedkYGoAQ&ved=0CD4Q6AEwAQ#v=onepage&q=monocle%20erotic&f=false

2. George L. Marshall Jr., "The Rise and Fall of the Newburgh Conspiracy: How General Washington and His Spectacles Saved the Republic." http://www.earlyamerica.com/review/fall97/wshngton.html

3. Pitt H. Herbert, "An Eye on the Monocle," *Optics Journal and Review of Optometry* 87, no. 7 (April 1 1950): 45–54.
4. Bruce J. W. Evans, "Monovision: A Review," *Ophthalmic and Physiological Optics* 27, no. 5 (August 17, 2007): 417–439. https://onlinelibrary.wiley.com/doi/full/10.1111/j.1475-1313.2007.00488.x
5. https://www.verywellhealth.com/what-is-monovision-3421638

7
Fringe Science

When people making laboratory glassware (or art glassware, for that matter) want to be certain that the piece they are working on is stress-free, they place it in a device consisting of two sheets of Polaroid, parallel to each other but separated by a distance, allowing the pieces to be inserted between them. The polarization axes of the two sheets are perpendicular to each other, and a diffuse light source is placed behind them. Normally, because the polarizers are crossed, the field of view is uniformly dark, but if something is placed between them that contributes a phase shift, it shows up instantly as a bright region. Stressed glass, which has a difference between the refractive indices along the direction of stress and perpendicular to it, shows up strikingly in a stress birefringence pattern. These devices can also be used for checking stress in mounted optics, in injection-molded plastic parts, and for identifying birefringent crystals.

A device like this in which one polarizer is fixed and the other rotatable—often with marked gradations—is called a "polariscope" or a "polarimeter." One in which the polarizers are fixed at right angles is often called a "colmascope," but this name is actually a trademarked device, first registered by American Optical Company in 1912 and renewed regularly since. (Although since the demise of American Optical the trademark has lapsed, effective January 20, 1997). The term "colmascope" has suffered the same fate as Kleenex and Cellophane, shifting from trademark to generic name. I still see old AO Colmascopes in labs and workshops.

But that bit of information—trademark registered in 1912—looks odd at first glance. Edwin Land patented his first type of Polaroid, the J-sheet polarizer, in 1929, and his improved H-sheet Polaroid in 1938. So what were they using as a polarizer in 1912?

It turns out that the polariscope actually has a long history, dating back to the 1830s. One of the most common designs used a pair of crossed Nicol prisms, which are made from the naturally birefringent crystal Iceland spar (a form of calcite). It was invented by William Nicol of Edinburgh (1770–1851). Light passing through a clear crystal of Iceland spar separates into the two different polarizations, but the separation is not large. Nicol found that by sawing a crystal along its short diagonal and cementing the two halves together with Canada balsam, one polarization was reflected at the interface, producing a much larger separation, or even effectively eliminating one polarization.

Because the Nicol prism polarizer is much thicker along the optical path than it is wide, a polariscope using crossed Nicol prisms cannot be a light and simple pair of sheets, like a modern device. It has to consist of two boxes, aligned along a common axis. And so sketches and photographs of these devices show that they are. They consist of two long boxes, with the long axes coincident.

In addition to their greater bulk, these polariscopes differed from modern devices in another interesting way. I first learned of it my first year in graduate school, when I obtained a copy of R. W. Wood's *Physical Optics* and read it all the way through. Wood was an amazing optical physicist, whose name is attached to a number of devices and phenomena. He was also a real "character," who used to walk the streets of Baltimore at night (he was a professor at Johns Hopkins), dressed in a black cloak and carrying bits of sodium wrapped in twists of tissue paper. He'd toss these into puddles of water, where they would burst into flame, astonishing passersby. He also wrote a book of comic verse, *How to Tell the Birds from the Flowers*, which he illustrated himself. The book is still available as an e-book.

Wood observed that the field of view seen through crossed Nicol prisms is not uniformly dark. If the field is large enough (and getting prisms large enough to provide a large field was a challenge), you could see that the dark region was actually a portion of a circular arc. There is actually a succession of these, alternating light and dark arcs. Wood called these "Landolt fringes" (other writers have called them "Landolt bands").[1]

I was fascinated by this, but the topic never came up in my optics classes or my reading. The phenomenon is observed in converging or diverging light, for the cases of crossed crystal polarizers, when the optic axis does not lie in the end face. Since crystal polarizers understandably fell out of use with the invention of sheet polarizers, this particular drawback in crystal polarimeters isn't an issue anymore.

This raises other questions: Who was Landolt? And why were the fringes named after him? A search through optical literature reveals very few uses of the term, and none predating Wood's use of it. I strongly suspect that Wood originated the term, and that all later uses trace back to his coining it in *Physical Optics* in 1883.

I had encountered the name "Landolt" with respect to those broken circles used in ophthalmic charts. Eye doctors use a chart having such Landolt rings in lines of different size and ask the patient being examined to tell in what direction the opening in the letter "C" is pointing. The chart was the invention of Edmund Landolt (1846–1926), a Swiss ophthalmologist who worked mainly in Paris. But why would an ophthalmologist be interested in polarization? And why couldn't I find anything about his fringes in anything I read about him?

Because, evidently, I had the wrong Landolt. The one I wanted was Hans Heinrich Landolt (1823–1910), a Swiss chemist who worked mainly in Aachen, Germany. His interest in polariscopes derived from his study of optical rotation. He was very careful in his measurements to measure precisely the degree of rotation, and he described their measurements in detail in the books and articles that he wrote about such polariscopes.

When using a Biot polariscope (also called a Mitscherlich polariscope), which is simply a fixed Nicol prism and a rotating Nicol prism fitted with a Vernier scale, he observed that the field of view would not be uniformly darkened when the polarizers were crossed, but there would be a dark fringe taking up a portion of the field. He urged those performing measurements to get this fringe centered in the field of view in order to obtain accurate results. Curiously, he invariably describes the fringe as a straight line, not as a portion of a circle, and illustrates it thus in his books. He himself did not claim to have discovered the phenomenon. He credited Austrian physicist Ferdinand Lippich with producing the explanation for the effect. The fringes could thus arguably have been called "Lippich fringes." Lippich himself took advantage of the fringe effect and used it to design a polarimeter that could measure optical rotation to within three arc seconds. (A later and fuller explanation was given by J. T. Groosmuller in 1926.) Wood evidently named them after the most common and well-known writer on the topic, and those quoting Wood followed suit.

The reason that the fringes assume the form of semicircles is that the direction of polarization is not uniform across the field of view. Because of obliquity factors, with rays not being completely parallel to the optical axes, the direction of polarization acts as if fanning out from a central point outside the field of view, rather than being parallel. When two such polarization fields are crossed, there is only a completely dark fringe at points where these fanned-out rays cross at right angles. It's easy to show, by geometrical construction, that the locus of these points of perpendicular intersection describe portions of circles. The same effect exists with some other birefringent polarizers, but it is not as pronounced (Figure 7.1).

Much of this explanation appears in the 1978 edition of the *Handbook of Optics*, in the chapter on polarization by Jean M. and Harold E. Bennett, but the extensive coverage in that edition has not carried over to later, much longer editions. This probably reflects the rarity of Nicol-based polarimeters and colmascopes today.

Finally, one last question. "Polariscope" and "Polarimeter" are clear enough, but why did American Optical come up with the name "Colmascope" to trademark? Certainly they wanted a name that would set their device apart from the other polariscopes on the market, so they cast about for something without "polari-" in the name. Their choice seems to derive from the Greek root *kolmo-*, meaning "perpendicular,"

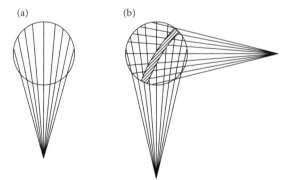

(a)   (b)

Figure 7.1 Illustration of how Landolt's fringes are formed. (a.) When viewed through a Nicol prism, the direction of polarization varies across the field of view, acting as if they originate from a point outside the field of view. (b.) When two Nicol prisms are crossed, the dark fringe only occurs when the directions of polarization are at right angles. The region where this occurs forms a semicircle.

Adapted from R. W. Wood's *Physical Optics*

since it has the two axes of polarization of the two polarizers set immovably perpendicular to each other.

Note

1. An example of following Wood's terminology: in 1908, F. E. Wright referred to the fringes as "Landolt's bands." This was three years after publication of Wood's *Physical Optics* in 1905. F. E. Wright, "The Bi-quartz Wedge Plate Applied to Polarimeters and Saccharimeters," *American Journal of Science* 176 (1908): 391–398.

References

Atkinson, Thomas G. *Oculo-Refractive Cyclopedia and Dictionary.* Entry on "Colmascope," p. 96. Chicago: Professional Press, 1921.

Driscoll, Walter G., and William Vaughn, eds. Chapter 10: "Polarization" by Jean M. Bennett and Harold E. Bennett. In *Handbook of Optics.* New York: McGraw-Hill, 1978

Groosmuller, J. T. "Das Polarisationsfeld Nicolscher Prismen," *Z. Instrumentenk* 46 (1926): 563.

Landolt, Hans. *Handbook of the Polariscope and Its Practical Applications.* Translated and adapted by D. C.. Robb and V. H. Veley. London: Macmillan and Co., 1882. https://archive.org/details/handbookpolaris01velegoog/

Landolt, Hans. *The Optical Rotating Power of Organic Substances and Its Practical Applications.* Translated by John H. Long. Easton, PA: The Chemical Publishing Company, 1902. https://books.google.com/books?id=wj793naWJcsC&pg=PA334&dq=Landolts+Fringes&hl=en&sa=X&ved=0ahUKEwjW_Nz278nUAhWDQiYKHYjSBYEQ6AEILjAB#v=onepage&q=Landolts%20Fringes&f=false

Lippich, F. "Über polaristrobometrische Methoden" *Wien. Sitzungsberichte Wiener Akademie* Band 91 (1885): 1059–1096.

Official Gazette of the United States Patent Office 176, no. 3 (March 19, 1912): 743. "Trademarks" section.

Wood, Casey A., ed. *The American Encyclopedia and Dictionary of Ophthalmology.* Entry on "Colmascope," vol. 4, p. 2343. Chicago: Cleveland Press, 1914.

Wood, Robert W. *Physical Optics.* 3rd ed. New York: MacMillan, 1934.

8
Light—as a Feather

Several pieces written on the history of the diffraction grating repeat a bit of information, often in precisely the same language (which indicates that many of them have copied from the same source), saying that the diffraction grating was discovered by James Gregory in the late 17th century, when he used a bird feather to split a beam of sunlight into different colors. Some sources go so far as to identify it as a feather from a seabird.[1]

> We can picture Gregory walking on our broad sands, watching the sea birds and idly picking up a feather and using it to discover a new phenomenon of light.

But this makes more, perhaps, than even Gregory himself intended.

James Gregory (November 1638—October 1675) was a Scottish mathematician, astronomer, and physicist. He served as professor at the University of St. Andrews in Fife, Scotland, and at the University of Edinburgh, and corresponded with the scientists of his day, most notably with Sir Isaac Newton. He is known for his work on infinite series, in which he derived series expansions for several trigonometric functions, and for his work on telescope design.

He corresponded heavily with John Collins (1624–1683), a publisher and accountant, with whom he discussed mathematical problems and physics, and whose correspondence is preserved today. The entirety of Gregory's work on the bird feather as a diffraction grating appears at the end of a letter discussing telescopes, dated May 13, 1673. It is an aside that Gregory suggested Collins bring up when he would next speak with Newton.[2]

> If ye think fit, ye may signify to Mr. Newton a small experiment, which (if he know it not already) may be worthy of his consideration. Let in the sun's light by a small hole to a darkened house, and at the hole place a feather, (the more delicate and white the better for this purpose,) and it shall direct to a white wall or paper opposite to it a number of small circles and ovals, (if I mistake them not,) whereof one is somewhat white, (to wit, the middle, which is opposite to the sun,) and all the rest severally coloured. I would gladly hear his thoughts of it.

Anyone who has worked with a transmission grating will instantly recognize the description of the undeviated zeroth order, along with the various orders on both sides of this, each showing separation of the constituent colors. Gregory did indeed use a bird feather as a diffraction grating, and he correctly stated that one with smaller separations between the barbs was preferable ("the more delicate . . . the better for this purpose"), which suggests that he experimented with several feathers, observing the effect.

There is no suggestion that he used a feather from a seabird. There is no indication that he, in fact, had any idea what was causing the effect, let alone that it was due to an interference effect that suggested the wave nature of light. One site reports that "The feather became the first diffraction grating but again Gregory's respect for Newton could have prevented him going further with this work."[3] But this is surely interpreting too much.

What we do know is that Gregory had observed a strange and very interesting effect, producing separation of colors in sunlight, just as Newton himself had done only a year before. But Gregory had used an ordinary, easily obtained bird's feather in place of Newton's fabricated glass prism. Furthermore, he had produced multiple spectra (as well as that unseparated, undeviated spot), in place of Newton's single spectrum. He would surely have thought Newton interested in this weird effect.

There is no indication that either of them carried on with this work. If Newton mentioned it in a subsequent letter, that letter has not survived, and is not among their correspondence. Gregory does not mention the incident in any of his later writings.

To the modern spectroscopist, that talk of circular and oval spots seems odd, accustomed as we are to the use of narrow line slits in our devices. Thomas Melvill observed the same thing when he first performed his flame spectroscopy in 1752. If either Gregory or Melvill had continued the work, he probably would have changed to slits to minimize overlapping of colors. But neither did. Both men died at early ages, Melvill at 27, and Gregory at 36. Even if they had intended to return to the experiments, their times had run out. And so the feather grating did not become a major experiment in the arsenal of the war between the wave and the corpuscular interpretation of light. It remained for Rittenhouse to consciously construct the first diffraction grating in 1786, and for Thomas Young to use early 19th-century COTS (commercial off-the-shelf) technology to perform the first measurement of the wavelengths of light in 1802.[4]

What influenced Gregory to try the feather experiment? Nothing in the earlier paragraph gives a clue. But the topic of bird feathers producing color had already been raised before Gregory's work. The natural philosopher Robert Hooke, in his book of explorations with his microscope, the 1665 book *Micrographia*, had commented on the colors observed in a peacock's feather. These, he observed, were due in some way to the structure of the feather, because dropping water onto the feather destroyed the vivid colors that had been observed.[5]

If Gregory knew of this work, he did not mention it in the portion quoted earlier, and one would think that the contrast between Hooke's colored peacock feather and his own white feather would be a cause for comment. One would expect Gregory to wonder, at least, why the peacock feather was brightly colored and his white feather was not. Or that it would lead him to suggest that, perhaps, the colors he observed might be due to the same effects, if differently manifested. Newton would later go on to write of the colors of the peacock feather in his *Opticks*, but he would not mention Gregory's experiment.

And so Gregory's experiment, like Claudius Ptolemy's measurement of the laws or refraction, Melvill's observations of flame spectra, and Rittenhouse's observations of diffracted orders, ultimately didn't lead anywhere, as they were not followed up, and it would not be until later that the principles were properly investigated and used to further our understanding. In this case, as in Melvill's, the scientist's untimely death may have contributed to the dropping of the study.

A modern investigation of the diffraction patterns produced by feathers was carried out in Spain in 2012 by Hugo Pérez García, Rafael García Molina, and Isabel Abril of the Universidad de Murcia and the Universitat d'Alacant. They measured the diffraction pattern obtained by shining a laser through a feather and compared it with the expected pattern, using measurements of the barb spacing obtained from electron microscope data. The results were in good agreement. Gregory would have approved.[6]

Notes

1. Accessed October 14, 2013. From the University of St. Andrews history site: http://www-history.mcs.st-andrews.ac.uk/history/Extras/Turnbull_address.html
2. Isaac Barrow (ed.), *Correspondence of Scientific Men of the Seventeenth Century*, vol. 2 (1841). Accessed October 14, 2013. http://books.google.com/books?id=0h45L_66bcYC&pg=PA254&dq=feather+ovals&output=html or http://books.google.com/books?id=P6ANAAAAQAAJ&pg=PA610-IA1&dq=Correspondence+of+Scientific+Men+of+the+Seventeenth+Century+...,+Volume+2&hl=en&sa=X&ei=A2ZcUoaLNffj4APIhIHoBg&ved=0CDkQ6AEwAQ#v=onepage&q=Correspondence%20of%20Scientific%20Men%20of%20the%20Seventeenth%20Century%20...%2C%20Volume%202&f=false The relevant quote is on p. 254.
3. http://www-groups.dcs.st-and.ac.uk/~history/PrintHT/U_of_St_Andrews_History.html
4. Stephen R. Wilk, *How the Ray Gun Got Its Zap* (New York: Oxford University Press, 2013). On Melvill, see 60–63; on Rittenhouse, see 57–59; on Young's measurement of light, see 181–185.
5. http://www.google.com/patents?id=ub1TAAAAEBAJ&printsec=abstract&zoom=4#v=onepage&q&f=false. See also Harold Baunhut's 1971 patent 3,592,533 for an "Optical Device for Simulating Optical Images." http://www.google.com/patents?id=zXp1AAAAEBAJ&pg=PA1#v=onepage&q&f=false
6. Hugo Pérez García, Rafael García Molina, and Isabel Abril, "Difracción de luz a través de una pluma de ave," *Revista Eureka Sobre Enseñanza y Divulgación de las Ciencias* 9, no. 1 (2012): 164–169. http://bohr.inf.um.es/miembros/rgm/TeachPubl/PerezGarcia_REurEDC9(2012)164-DifraccionPluma.pdf

9

Lacemaker's Lamps

How did artisans doing fine work illuminate their work before the invention of modern lighting? The obvious answer is that they used daylight. Scriptoria in the Middle Ages had large windows that illuminated the manuscripts. Fine filigree and mosaic work around the world was presumably done with plenty of bright daylight.

Using artificial light during night-time, especially during the long nights in Northern Europe, would allow practitioners to extend their work hours and bring in more money. But premium bright and smokeless light sources, like beeswax and spermaceti candles, were expensive. Oil lamps, rushlights, and tallow candles produced less light and light not quite so stable (and, after the 17th century, candles in Britain were taxed). So what could an artisan making fine detail work, such as lace, use?

One option was to place a lens in front of a candle or lamp, or to place a curved mirror behind it to approximately collimate the light. I have, in fact, seen such an antique fixture for a candle having a thick lens in front of it, which is what inspired this column. But both lens and mirrors would probably have been too expensive for most lacemakers—thick pieces of decent optical-quality glass were not easy to come by, nor were good-quality curved mirrors. A much more economical option was to place a hollow glass globe filled with water near the candle.

This is much like the solution described by Seneca for use in Imperial Rome—a glass sphere that could be filled with water would provide a good short-focal-length lens of better quality than most of the glass lenses of the day would give, and it was easy to produce. If the glass ball got too cloudy with growing algae, it could be emptied and washed. But most users avoided this by using freshly gathered "snow water," the poor man's distilled water ("The purest of all the common waters" one 1815 treatise called it[1]) and tightly stoppering the container. Seneca used such globes for magnification, to aid his failing eyesight, but such water bulbs could be used as light concentrators as well.

Sometimes the water was tinted blue. The globe could be supported between the worker and the light source by a stand or hung from the ceiling. There exist several antique candle stools which have a central candle or lamp, surrounded by holes into which the water-filled glass bulbs were placed.[2] Some works describe many workers clustered around a single candle, each providing his or her own bulb from a protective cloth bag, trying to maximize the number of people who could gain an advantage from a single source of light:[3]

> In the evenings eighteen girls worked by one tallow candle, value one penny; the "candle-stool" stood about as high as an ordinary table with four legs. In the middle of this was what was known as the "pole-board," with six holes in a circle and one in the centre. In the centre hole was a long stick with a socket for the candle at one end and peg-holes through the sides, so that it could be raised and lowered at will. In the other six holes were placed pieces of wood hollowed out like a cup, and into each of

these was placed a bottle made of very thin glass and filled with water. These bottles acted as strong condensers or lenses, and the eighteen girls sat around the table, three to each bottle, their stools being upon different levels, the highest near the bottle, which threw the light down upon the work like a burning glass.

These devices were not limited to lacemakers, as the various names for them in different places indicate. Brian Lemin, in an excellent internet article on them, says that in Denmark they were called "cobbler's lamps." In Germany they are *Schusterkugel*, "shoemaker's lamps," and were used by goldsmiths and theaters as well.[4]

One of the more interesting variations is that a globe filled with salt water was used in the 17th century by Robert Hooke to illuminate the sample in his pioneering microscope. Hooke provides an illustration of his illuminator, which consisted of a lamp, the brine-filled globe, and a rather thick plano-convex glass lens in his ground-breaking book *Micrographia* (1665).[5] He does not, however, explain the reason that he constructed it this way. Did Hooke simply adopt a common condenser in use by the tradesmen of his time, in addition to the glass lens? Granted that the glass available to Hooke was not of high optical quality (while a water-filled sphere was), why did Hooke use *both* a lens and a globe filled with salt water?

Could there have been some reason other than simply focusing the light? Hooke was concentrating light from a flame onto a small patch, and he would not want to cook his subject. A water-filled globe will not only concentrate light, it will absorb the infrared portions of the light as well. You can purchase water-filled condenser filters for microscopes today that use distilled water to eliminate infrared for precisely this reason.

Of course, infrared light was not discovered until William Herschel discovered "calorific rays" in 1800. But perhaps Hooke found that he had better results with a water-filled globe in his magnifier chain than using only a lens. He might have observed that his samples did not get as hot or deteriorate as rapidly. And perhaps those lacemakers, cobblers, and goldsmiths with their globes of water, often colored blue,[6] found that it strained their eyes less, even if they did not know the reason.[7,8,9,10]

In this light, it is particularly interesting to read an excerpt on the use of such water-filled globes for illumination from John Jackson's *A Treatise on Wood Engraving, Historical and Practical*, printed in London in 1839. The book is illustrated by more than 300 illustrations, engraved, of course, in wood, by the author himself. He writes:[11]

> by use of these globes one lamp will suffice for three or four persons, and each person will have a much clearer *and cooler light* than if he had a lamp without a globe solely to himself. (emphasis added)

Notes

1. Charles Hutton, *A philosophical and mathematical dictionary: containing an explanation of the terms, and an account of the several subjects, comprised under the heads mathematics, astronomy, and philosophy both natural and experimental; with an historical account of the*

rise, progress and present state of these sciences; also memoirs of the lives and writings of the most eminent authors, both ancient and modern, who by their discoveries or improvements have contributed to the advancement of them. Vol. 2, entry on "Snow-water" (1815).

2. Brian Lemin reproduces images of many of these in his paper "The Great Deception – Lace Makers Lamps" (2010, available at several Internet sites: http://www.cs.arizona.edu/patterns/weaving/webdocs/lb_lamps.pdf). The title of his paper refers to the bulbous glass-bulb lamps called "lacemaker lamps" in the antique trade, which do not follow the quoted description, and in which the glass bulb appears to simply be a fuel reservoir, not a magnifier.

3. Mrs. Roberts of Spratton in England. Cited by Mary Carolyn Beaudry in *Findings: The Material Culture of Needlework and Sewing*, 157.

4. http://www.pepysdiary.com/diary/1664/10/07/

5. https://images.search.yahoo.com/images/view;_ylt=AwrB8pEkO1hUzVgA.X6JzbkF;_ylu=X3oDMTIycm51bTVvBHNlYwNzcgRzbGsDaW1nBG9pZAM5MDc4MjU3OTJjZTUwMjg1ZGE1MGNlM2E0YzhmNGNlZgRncG9zAzIEaXQDYmluZw-- ?back=https%3A%2F%2Fimages.search.yahoo.com%2Fsearch%2Fimages%3F_adv_prop%3Dimage%26va%3DHooke%2527s%2Bmicroscope%2Billustration%26fr%3Dyfp-t-901%26tab%3Dorganic%26ri%3D2&w=1280&h=1415&imgurl=25.media.tumblr.com%2Ftumblr_m8fjxtKYoO1qgzqeto1_1280.png&rurl=http%3A%2F%2Fscientificillustration.tumblr.com%2Fpost%2F30261587945%2Frobert-hookes-microscope-as-shown-in&size=374.8KB&name=tumblr_m8fjxtKYoO1qgzqeto1_1280.png&p=Hooke%27s+microscope+illustration&oid=907825792ce50285da50ce3a4c8f4cef&fr2=&fr=yfp-t-901&tt=tumblr_m8fjxtKYoO1qgzqeto1_1280.png&b=0&ni=160&no=2&ts= &tab=organic&sigr= 12u0cfrmu&sigb=1401u439n&sigi= 11n00efeg&sigt= 113s680co&sign= 113s680co&. crumb=zGDeO0tF5V0&fr=yfp-t-901

6. Blue colorants absorb or scatter red light, and this can easily carry over into the near infrared as well. The windows in the Visitor's Gallery at the Laboratory for Laser Energetics at the University of Rochester, back when the system operated at 1.06 microns, consisted of two glass plates with blue copper sulphate solution between them to absorb stray laser reflections.

7. Ed Welch, "Researching a Unfamiliar Object," accessed September 2014, http://www.journalofantiques.com/May04/businessmay04.htm

8. "FlashStools," http://www.mkheritage.co.uk/cnm/html/EXHIBITS/lace/lacehtml/13_flashstool.html

9. Gertrude Whiting, *Old Time Tools and Toys of Needlework* (1928), 253.

10. Robert B. Graves, *Lighting the Shakespearean Stage 1567–1642* (1999), 24.

11. John Jackson, *A Treatise on Wood Engraving, Historical and Practical* (London: Charles Knight and Co., 1839), 652. Accessed January 20, 2015, https://books.google.com/books?id=umFdAAAAcAAJ&pg=PA652&dq=%22bull%27s+eye%22&hl=en&sa=X&ei=aKGuVM2LEeTksATt5oHwBw&ved=0CDkQ6AEwAzgK%20-%20v=onepage&q=%22bull's%20eye%22&f=false#v=snippet&q=%22bull's%20eye%22&f=false

10
Thoreau's Rainbow

Once it chanced that I stood in the very abutment of a rainbow's arch, which filled the lower stratum of the atmosphere, tinging the grass and leaves around, and dazzling me as if I looked through colored crystal. It was a lake of rainbow light, in which, for a short while, I lived like a dolphin. If it had lasted longer it might have tinged my enjoyments and life.
—Henry David Thoreau, *Walden, or Life in the Woods* (1854)

Recently, I finally got around to reading Henry David Thoreau's classic book *Walden: Or A Life in the Woods*, in the annotated edition with notes by Philip Van Doren Stern.[1] The book is often seen as a paean to the simple life and a celebration of nature. The book is that, and much more, a complex and enduring work. Thoreau's observations of nature abound in the book, and his modern editor praises him for his accuracy. "He did not pretend to be a professional naturalist," notes Van Doren (p. 6). "Although he was a perceptive observer and a first-rate note-taker, far better than most professionals in the field, then and now." So when the perceptive and accurate Thoreau writes that he stood at the proverbial end of the rainbow, one tends to take notice. One who did notice and criticized Thoreau for it in the pages of the *Atlantic Monthly* was heavily attacked as an artless philistine by pro-Thoreau readers. The critic was naturalist and writer John Burroughs (1837–1921), who was already stirred up controversy in the *Atlantic* for his criticism of "nature fakers," authors such as Ernest Thomson Seton whose depictions of animal behavior strayed, in Burroughs's opinion, too far from reality. There was a lively controversy over it in 1903–1905,[2] but it was undoubtedly recalled in 1920 when Burroughs applied the same critical sense to Thoreau's *Walden*.

As an optical engineer, and one who has lectured about the rainbow frequently, I can't help but agree with Burroughs's comments. He properly points out that the rainbow is a phenomenon that must exist at an angle of 42° from the antisolar point and is quite impossible to approach. Those who wrote in to him claimed to have approached rainbows themselves. But Burroughs observed that "No one has written me that he stood in the abutment of a rainbow yesterday or today. It is always on some occasion long ago."[3] Imperfect recollection, in other words, perhaps colored by expectation.

That the end of the rainbow is unattainable is enshrined in at least two pieces of folklore. One is the familiar idea that there is a pot of gold at the end of the rainbow. (Although everyone agrees that this is an Irish belief, I'm having a hard time pinning it down. Several Internet sources claim that it dates back at least to the 17th century, but cite no sources. I have found references to a silver cup at the end of the rainbow dating from 1833.[4]) A Bulgarian folk belief is that anyone who passes under the rainbow changes gender.[5] But this feature of the rainbow was apparently not evident to Burroughs's critics.

"Just how Thoreau deluded himself, I am at a loss to know," wrote Burroughs. Later he wrote, "How Thoreau ever found himself standing in the bow's abutment will always remain a puzzle to me." Van Doren Stern, ever championing Thoreau, cites an article by Charles D. Stewart, "A Word for Thoreau," in rebuttal.[6] "As one who has studied the rainbow both at Niagara Falls and with the garden hose, I will venture to say that Thoreau was telling the truth," Stewart wrote. Nothing in the descriptions that follow, however, indicates that Stewart ever actually reached a rainbow. They were in profusion around him, and appeared at times to be close at hand, but he never does say that he actually reached the foot of one. "I am quite prepared to take my stand with Thoreau," he wrote. "If a low fog or wisp of vapor came floating along I do not doubt that Thoreau could have seen the foot of the rainbow right in front of him."

Well, not fog or a wisp of vapor—fogbows are notoriously white and uncolored—but a local sprinkling of water, as from a lawn sprinkler, could give the impression of the rainbow being nearby. But this doesn't match Thoreau's description—look at the epigraph earlier. "Tinging the grass and leaves around ... It was a lake of rainbow light, in which, for a short while, I lived like a dolphin." [7] That does not sound as if Thoreau could see a spot where the rainbow struck the ground before him—it reads as if he has been plunged among the colors of the rainbow, with them surrounding him, rather like the *putti* and *pegasi* shown in the Pastoral Symphony segment of the Disney film *Fantasia*. And it is to this that Burroughs objected.

A more nuanced apology for Thoreau appeared shortly after this in the pages of *The Journal of the Royal Astronomical Society of Canada*. [8] "Did Thoreau really intend to state that he actually had been within a rainbow which he himself could recognize as such?" asks C. A. Chant. "I do not believe that he did." He goes to imagine a plausible scenario. "On the occasion referred to by him there were present a bright sun and a mist consisting of fine water drops. From them there resulted an iridescence which enveloped him and which he felt must be like that in the end of a rainbow—which would be seen as such by someone in the proper position."

A modern take with a similar interpretation appears in the blog *The Curious People* by David Bristow for November 27, 2014. [9] "Who knows what Henry experienced. I suspect that he's using hyperbole here and that he assumed we wouldn't take him literally when he's clearly stating what we all know is a physical impossibility. He was prone to doing that for effect, and I think what he's trying to convey here is the absolutely glorious aspect of the light. It was *as if* he were standing inside a rainbow. I think that would be consistent with his style."

Certainly Thoreau was no naïve and simple writer. Throughout *Walden* he takes incidents recorded in his *Journal* and rearranges them or tweaks them for effect. This is very clear in Van Doren Stern's annotated edition, which points out such deviations. And, as van Doren Stern observes, "His writings on natural history belong to art rather than to science" (p. 6). Relevant to this is a quote from Thoreau's journal of November 5, 1857, quoted in Bristow's blog: "I think that the man of science makes this mistake ... that you should coolly give your chief attention to the phenomenon which excites you as something independent of you, and not as it is related to you. The important fact is its effect on me. He thinks that I have no business to see anything else but just what he defines the rainbow to be." [10]

The problem with this attitude—that those who don't see the art, and insist on scientific literalism, are simply *nyekulturny* philistines who have no soul—is that their art is not the only game in town, and scientific accuracy does have its place, even in art. And if one is going to make much of Thoreau's wonderful skill at observation and accuracy in his descriptions of the natural world elsewhere in *Walden*, one can't complain about someone else pointing out errors in it when he describes the rainbow. Saying that the fact that he's describing a literal impossibility, which ought to clue readers in on the fact that he has taken off in a flight of fancy, is a poor excuse when you realize that most people don't realize that this *is* an impossibility.

There's an echo here of the controversy over John Keats's 1820 poem "Lamia," in which he takes Isaac Newton to task for mathematically explaining the rainbow. From lines 229–238 of Part II of the poem:

> Do not all charms fly
> At the mere touch of cold philosophy?
> There was an awful rainbow once in heaven:
> We know her woof, her texture; she is given
> In the dull catalogue of common things.
> Philosophy will clip an Angel's wings,
> Conquer all mysteries by rule and line,
> Empty the haunted air, and gnomèd mine—
> Unweave a rainbow, as it erewhile made
> The tender-person'd Lamia melt into a shade.

Keats's complaint has been answered by many scientists. Richard Dawkins even entitled one book *Unweaving the Rainbow*, [11] and he devoted much space to the scientist's argument that, far from robbing the rainbow of its glory and awe, science reveals more aspects of the phenomenon to excite our admiration.

Notes

1. Henry D. Thoreau, *The Annotated Walden; or, Life in the Woods*, edited and annotated by Philip Van Doren Stern (New York: Bramhall House, 1970). Quotation on p. 332, Chapter 10. One thing I learn from this edition is that I have been pronouncing the author's name incorrectly. It should rhyme with the last name of Edward Murrow.
2. Wikipedia, "The Nature Fakers Controversy," accessed June 21, 2016. https://en.wikipedia.org/wiki/Nature_fakers_controversy
3. John Burroughs, "The Unapproachable Rainbow," *Atlantic Monthly* 126, no. 1 (July 1920): 98–101.
4. Elizur Wright Jr., *The Sin of Slavery and Its Remedy* (New York, 1833, printed by the author), 35. https://www.google.com/books/edition/The_sin_of_slavery_and_its_remedy/tNdt93_23vIC?hl=en&gbpv=1&dq=%22end+of+the+rainbow%22&pg=PA35&printsec=frontcover#spf=1569897619579

5. Carl Boyer, *The Rainbow from Myth to Mathematics* (Princeton, NJ: Princeton University Press, 1959, reprinted 1987). Also cited—without giving the sources—by Wikipedia. https://en.wikipedia.org/wiki/Rainbows_in_mythology

6. Charles D. Stewart, "A Word for Thoreau," *Atlantic Monthly* 156, no. 1 (July 1935): 110–116. Stewart critiques Burroughs's writing in "the last book he ever wrote," not on Burroughs's 1920 article.

7. This is undoubtedly Thoreau's broad knowledge giving rise to a witticism. Not only is he suggesting that he swims in a sea of colored light, as a dolphin swims in the sea, but it also recalls the saying about the "dying colors of the dolphin." The fish called the dolphin (*Coryphaena hippurus*, also called the mahi mahi or the dorado) changes colors as it dies, something much written of in Thoreau's day.

8. C. A. Chant, "Thoreau and His Rainbow," *Journal of the Royal Astronomical Society of Canada* 29, no. 6 (July–August 1935): 225–227. Accessed June 21, 2016, http://articles. adsabs.harvard.edu//full/1935JRASC..29..225C/0000227.000.html

9. Bristow, David blog "The Curious People" Accessed June 21, 2016, https://thecuriouspeople. wordpress.com/2014/11/27/standing-inside-a-rainbow-walden-145/

10. Bristow, David, *op. cit.*, who, in turn, got it from Robert Richardson and Barry Moser, *Henry Thoreau: A Life of the Mind* (Berkeley and Los Angeles: University of California Press, 1988), 363.

11. Richard Dawkins, *Unweaving the Rainbow: Science, Delusion, and the Appetite for Wonder* (Boston: Houghton-Mifflin, 1998).

11

The Well-Tempered Spectrometer

Color is the keyboard, the eyes are the harmonies, the soul is the piano with many strings. The artist is the hand that plays, touching one key or another, to cause vibrations in the soul.

—Wassily Kandinsky (1866–1944)

Musical tones occur as a continuous run of frequencies or wavelengths, varying smoothly across the range of human hearing. Seen in this way, our system of dividing up the tones into notes and octaves is highly artificial. There's no fundamental physical reason to have octaves or semitones the way we do. We could, in principle, play all our music on instruments like the theremin, the trombone, and the slide whistle and dispense with notes altogether.

We don't, of course, for a number of reasons. One is that not all instruments lend themselves to the production of a truly continuous range of tones, and they can only produce certain wavelengths due to the physics of their construction. More important is the way the human mind and brain respond to combinations of tones. We like combinations in which the frequencies are integral multiples of each other, or where the ratio of frequencies is 2/3 (a perfect fifth) or 3/5. But we dislike many other pairs of frequencies played simultaneously, branding them as "dissonant." Exactly why this should be so has been the subject of much speculation, some of it tongue in cheek.

Since most instruments are constructed with fixed tones, and these must be tuned to produce a pleasing sound, this issue of the proper tuning of lyres, pianos, horns, and other instruments is of great importance to serious musicians, composers, and music theorists. Most of the rest of us simply take it for granted, assuming that simply saying that we have the notes *do, re, mi,* and so on (along with associated sharps and flats) as sung to us by Julie Andrews or Mary Martin explains it all. But it doesn't, by a long shot. There is a large and, to the nonexpert, bewildering body of knowledge and experience involved in the tuning of instruments and the construction of musical scales required to make things sound euphonious.

For most practical purposes, and for the purposes of this discussion, there are two main methods of importance. One of these is Pythagorean tuning, in which the ratios of the frequencies (or wavelengths), and thus of, say, the strings in a lyre, form simple ratios, most of them 3:5. This was relatively easy to achieve, and it produced pleasing sounding notes and combinations for most uses. But there were some cases where what should be a fifth ended up out of its proper ratio, and the resulting "wolf" fifth sounded harsh. Extending the system across several octaves was problematic as well. The notes G-sharp and A-flat, which in a simplified system of half-steps ought to be equivalent, in fact turn out to be significantly different.

The solution to this was to "smear out" the errors among all the notes in a scale by making half-tone steps of truly equal size. This is called "equal temperament" (more precisely, "twelve-tone equal temperament," abbreviated 12-TET or 12-ET). The frequency interval between adjacent notes (including sharps and flats) is the twelfth root of two ($2^{1/12}$), making them logarithmically equal. The invention of equal or even temperament, it has been argued, occurred very nearly simultaneously in China and in Europe, by Zhu Zaiyu (also rendered as *Chu-Tsaiyu*) and by Simon Stevin (also known by his Latin cognomen *Stevinus*) around 1584 (which is just before the invention of logarithms).

Equal temperament is by far the most widely used musical system today, although it took a long time for it to become generally accepted. It allows the playing of harmony and chords with great flexibility and, to the ear accustomed to it (as the modern ear is—we have been "trained" to hear this as the normal scale), euphonious tones across all octaves and all instruments. Johann Sebastian Bach wrote his famous *The Well-Tempered Clavier* in 1722 as a sort of propaganda piece to show the utility of a well-tempered scale. (Although, it should be noted, that "well-tempered" means one in which music played in most of the major and minor keys doesn't sound dissonant. It includes equal temperament but isn't restricted to it.)

Why bring all of this up in an essay that is supposed to be about optics? Because what I said earlier about music is equally applicable to color. Colors occur as a continuous run of frequencies or wavelengths, varying smoothly across the range of human vision. Seen in this way, our system of dividing up the spectrum into individual named colors is highly artificial. There's no fundamental physical reason to have fixed colors the way we do.

There have been various divisions of the rainbow into colors, and different identifications of what those colors are, by various people over the years. But our current division into seven colors of Red, Orange, Yellow, Green, Blue, Indigo, and Violet is due to Isaac Newton, who gave his assessment of the situation in his book *Opticks* (1704), in the first part of the second book.

I have written about this at length before,[1] but suffice it to say that Newton saw the analogy between colors and musical notes, and he used the analogy to associate each note with a different color. Thus, since there are seven distinct notes in an octave (ignoring the octave itself, and taking no notice of the sharps and flats), Newton decided that there should be seven colors. Furthermore, even though Newton was opposed to the wave theory of light, he knew there were characteristic lengths associated with the colors, and the proportions of these lengths, he felt, ought to correspond to the ratios of wavelengths in the notes. The fact that Violet light has a characteristic length just about half that of extreme Red must have been a sort of confirmation to Newton of his hypothesis. (The identification of colors with notes occurs when Newton is describing the phenomenon we today call "Newton's rings," in which we see white light interference between light reflected from a flat glass plate and a section of a glass sphere. Because the separation between the sphere and the plate it is resting on—the *sag*—can be calculated mathematically, and because there are individual colors at each *sag* distance, Newton could identify this characteristic length with each color. The fact that the spectrum repeated several times with increasing *sag*, as the scales of a piano repeat, and the fact that the *sags* for appearances of the same color were in integral proportion with each other, as the octaves are, were probably responsible for suggesting the analogy to Newton. He basically followed through by drawing other analogies between the two situations, including identifying colors with notes. When Thomas Young, champion of the wave theory of light, first measured the wavelengths of different colors of light a century later, he had Newton's numbers to compare with his own.)

But what length or frequency ratios did Newton use? It was a live issue in Newton's day. Would he use Pythagorean temperament, with its commitment to pleasing harmony, or equal temperament, with its commitment to reproducibility across octaves? In fact, he used neither, but his choice was closer to Pythagorean. Newton used yet another method, that of *just temperament*, in which the ratios of wavelengths are given by ratios of whole numbers, as in the Pythagorean, but which keeps the numbers used in those ratios small. In the Pythagorean tuning, for instance, the note B is 128/243 (= 0.5267 …), which is the ratio of two relatively large numbers. In just intonation, the ratio is the much simpler 8/15 (= 0.5333 …).

Because of his use of tone as an analogy for color, Newton was forced to extend his original five-color spectrum to seven. It must have been obvious to place Orange (which would be D, or *re*) between Red (C, or *do*) and Yellow (E, or *mi*), but it wasn't clear what to use for A or *la*, between Blue and Violet. Newton eventually settled on Indigo and argued that you could perceive this distinct color between Blue and Violet.

All of this has an element of arbitrariness in it—why just intonation rather than Pythagorean or equal tempering? Or why not some other system? In fact, why stick to only eight tones? Why not include all semitones in the scale—the sharps and flats, the "black keys" on a piano—and have an "octave" consisting of twelve distinct colors, rather than only seven. "Twelve" lacks the mystical associations of "seven," but it would give us greater resolution in describing colors.

The point here is that our system of seven colors in the rainbow is no more a "given" than the choice of musical scale is. In fact, it's less "natural," being that there are no dissonances in color mixing as there are in musical chords. Two musical notes close together produce an unwanted "beat" effect, but mixing two nearby colors simply produces an average tone. And there are no "wolf" fifths in color mixing.

So what *would* our color spectra look like if Newton had decided to use Pythagorean temperament, or equal temperament, or a full twelve-tone chromatic scale with equal temperament (in which the "colors" rise in the ratio $2^{-(n/12)}$)? I have constructed some tables, using the ratios Newton did, modern "proper intonation" values, representative values of colors, and an even-tempered scale. In all cases, I set Yellow to 580 nm and measure values from there. Yellow as a color occupies a very narrow range, so it seemed a good place to use as a reference. It's also about in the center of the visual scale of colors.

Newton's Tempered Values

Color	Ratio	Wavelength (nm)
Red	1/1 = 1	696
Orange	8/9 = 0.8889	618.67
Yellow	5/6 = 0.8333	580

Color	Ratio	Wavelength (nm)
Green	3/4 = 0.75	522
Blue	2/3 = 0.6667	464
Indigo	3/5 = 0.6	417.6
Violet	1/2 = 0.5	348

The Values of Colors Given Today by Such Sources as Wikipedia

Color	Ratio	Wavelength (nm)
Red	1.00	700
Orange	0.886	620
Yellow	0.829	580
Green	0.757	530
Blue	0.671	470
Indigo	0.643	450
Violet	0.6	420

The agreement is actually pretty good, only departing significantly for Indigo and Violet, especially considering that the modern color definitions are pretty arbitrary.

Modern Just Intonation Values

Color	Ratio	Wavelength (nm)
Red	1/1 = 1	734.1
Orange	8/9 = 0.8889	652.5
Yellow	64/81 = 0.790	580
Green	3/4 = 0.75	550.5
Blue	2/3 = 0.6667	489.4
Indigo	16/27 = 0.593	386.7
Violet	128/243 = 0.527	367.0

Modern Equal-Tempered Values

Color	Ratio	Wavelength (nm)
Red	1	730.8
Orange	$2^{-(2/12)} = 0.891$	651.0
Yellow	$2^{-(4/12)} = 0.794$	580
Green	$2^{-(5/12)} = 0.749$	547.4
Blue	$2^{-(7/12)} = 0.667$	487.7
Indigo	$2^{-(9/12)} = 0.595$	434.5
Violet	$2^{-(11/12)} = 0.530$	387.1

There's not actually an enormous difference between these scales, if you compare the spectral colors. Where the differences are largest, the colors are fairly broad, and the fit forgiving.

So what about taking it to the full extent—what if we had an even-tempered full chromatic scale, with all the sharps and flats, the black keys as well as the white on our scale, as they are on the piano?

Modern Equal-Tempered Values

Color	Ratio	Wavelength (nm)
Red	1	730.8
?	$2^{-(1/12)} = 0.944$	689.7
Orange	$2^{-(2/12)} = 0.891$	651.0
?	$2^{-(3/12)} = 0.841$	614.5
Yellow	$2^{-(4/12)} = 0.794$	580
Green	$2^{-(5/12)} = 0.749$	547.4
?	$2^{-(6/12)} = 0.707$	516.7
Blue	$2^{-(7/12)} = 0.667$	487.7
?	$2^{-(8/12)} = 0.630$	460.3
Indigo	$2^{-(9/12)} = 0.595$	434.5
?	$2^{-(10/12)} = 0.561$	410.1
Violet	$2^{-(11/12)} = 0.530$	387.1

What names would we give these new "colors"? Imagine yourself in the role of Newton, trying to convince your friends that there were obvious and distinct colors between our currently recognized ones, and trying to come up with descriptive names for them. Probably, like most color namers, we would be thrown on the necessity of naming them by analogy, as the color *Orange* is named after the fruit, or as Newton named *Indigo* after the dye plant. Would the new Red-Orange be *Watermelon*? Could the Orange-Yellow be *Mango*? An easy way out would be to christen Green-Blue as *Aqua*. But Blue and Indigo are, I confess, so similar to me already that trying to split the difference with an appropriate descriptive hue seems difficult. I would probably take refuge in *Royal Blue* or *Prussian Blue* or *Midnight Blue*.

Note

1. See "ROY G BIV," *The Spectrograph* 21, no. 1 (Fall 2004): 9–10, or as chapter 7 (pp. 47–52) in my *How the Ray Gun Got Its Zap* (New York: Oxford University Press, 2013).

12

Sacred Sun

In 1920, the state of New Mexico, which had only been one of the United States for eight years, needed a flag. The Daughters of the American Revolution sponsored a contest for the new flag. The winning entry was submitted by Dr. Harry Mera, a physician and an anthropologist at the Museum of Anthropology in Santa Fe. His design was a very simple one, inspired by a symbol on a pot made by an anonymous potter from the Zia Pueblo. The symbol was the Zia sun symbol, a circle with sixteen rays extending from it. The rays did not radiate from the center of the circle, but consisted of four groups of four parallel rays, each with two long central rays flanked by two shorter parallel rays, with all rays either horizontal or vertical. The original pottery symbol had a stylized face, consisting of two triangular eyes and a rectangular mouth, but Mera left these features out, opting for the bold outline alone. The sun symbol was executed in red/maroon on a yellow background. These were the colors of the imperial Spanish flag, so that the flag combined elements of the state's indigenous inhabitants and of the Spanish heritage. Mera's wife helped to fabricate the flag.

At first, there were complaints about the "heathen" imagery of the flag, but over time it became a beloved image. A 2001 survey by the North American Vexillological Association pronounced it the best-designed flag of any US or Canadian state or territory. A *USA Today* poll agreed. The symbol found its way onto the New Mexico license plate and official state documents, and it has been adopted by tourist sites and businesses across the state.

The problem is that no one asked the people of Zia Pueblo about this. The Zia, a private people, ask visitors to refrain from taking photographs or even making sketches of ceremonies. The Zia sun symbol is one of the group's oldest motifs, dating back to at least 1200 CE, and they feel that it has been improperly appropriated from them, and that its use on motels, motorcycles, portable toilets, and the like "dilutes its sacred meaning and disparages the Zia people." They contend that the piece that served as Dr. Mera's inspiration must have been stolen, since only ceremonial pottery would have had the symbol, and these would not be allowed to leave the Pueblo. (The pot was repatriated to the Zia Pueblo kiva in 2000.) Legislation regarding the ownership, use, and meaning of the symbol was being debated in the New Mexico state senate in 2014. That same year the National Congress of American Indians passed a resolution recognizing the Zia Pueblo's cultural property right to the symbol.[1] In 2018, Zia Pueblo governor Anthony Delgarito presented the argument for protecting the symbol at a meeting of the World Intellectual Property Organization (WIPO) in Geneva, Switzerland.[2]

The official flag of the Zia Pueblo features the symbol, but in red on a white background, rather than the colors of the old Spanish flag (Figure 12.1).

The appeal and meaning of the Zia sun symbol are clearly of deep importance, and I certainly do not mean to belittle that or to minimize its meaning. But I would suggest

Figure 12.1 The flag of the Zia Pueblo, showing the proper colors.
Reproduced with the permission of the Zia Pueblo.

that the symbol might itself have been inspired by a not wholly uncommon optical phenomenon, and that the same phenomenon inspired religious symbols of the Jews, the Christians, and prehistoric Europeans.

I have been aware for some time that striking effects can be produced by sunlight reflected and refracted by ice crystals in the atmosphere. These crystals generally take one of two basic forms. They can be thin hexagonal plates with parallel surfaces. Or, reversing the aspect ratio, they can be long, thin "needles" with hexagonal cross sections. Occasionally the plates or needles will have an equilateral triangle cross section. Reflection and refraction from such crystals can produce a wide range of haloes and arcs around the sun. Among the more familiar are haloes around the sun, separated by 22° and 46°, caused by light passing through the plates and refracting from surfaces with, respectively, a 60° or a 90° vertex angle. This demands that the crystals be randomly oriented in the sky, and this is often the case with cirrus clouds, or if one is near snow-making machinery. Large hexagonal plates tend to orient when falling in relatively still air, with the hexagonal cross section horizontal. This produces the phenomenon of "parhelia," or "sun dogs"—bright spots 22° on either side of the sun. Many other effects can also be observed when conditions are right, but these are often subdued, and people do not expect them, as they do the rainbow, and so their existence goes unnoticed most of the time. Only when a really spectacular display is seen do people take notice and wonder.

There are two effects that do not rely (or at least, do not chiefly rely) on internal refraction through the ice crystal. One of these is the solar pillar, which is a band of light extending above and below the sun. It is caused by sunlight reflecting from the broad, flat, horizontal faces of horizontally oriented hexagonal ice plates.

Another feature that can be seen when ice crystals are oriented horizontally is the parhelic circle, which appears to be at the same altitude as the sun and makes a horizontal line. Ideally, it can produce a circle around the entire sky, seeming to be at the same altitude, but in reality the needed crystals often only cover a small portion of the sky. The effect is caused by light reflecting from the vertically aligned faces of those horizontally oriented plates. Some rays, especially away from the sun, can be caused by light entering the plates through a vertical facet, reflecting from other vertically oriented faces, then exiting. These rays can have some color separation, and they can give a slight tinge to the parhelic circle (Figures 12.2, 12.3).

Figure 12.2 Setting sun with haloes and features formed by light reflecting and refraction from hexagonal ice crystals in the atmosphere. The sun is surrounded by the 22° halo, flanked on right and left by parhelia (or sun dogs). Extending vertically above and below the sun is the sun pillar, and extending horizontally through the sun is the parhelic circle.
Shutterstock

The result of these two effects of reflection from horizontally oriented hexagonal ice plates is to give both a vertical and a horizontal line through the sun, creating a "cross." It looks, in fact, very much like the Zia sun symbol, although there is only a single, if wide, beam going in each of the four cardinal directions of Up, Down, Left, and Right, not four in each of these. The appearance is strikingly similar.

If the idea of an image of deep religious significance might be inspired by a striking natural phenomenon, then consider something from the Judeo-Christian tradition—Ezekiel's wheel. In the first chapter of the book of Ezekiel, the sixth-century BCE prophet writing from the exile in Babylon:

1. Now it came to pass in the thirtieth year, in the fourth *month*, in the fifth *day* of the month, as I *was* among the captives by the river of Chebar, *that* the heavens were opened, and I saw visions of God.
2. In the fifth *day* of the month, which *was* the fifth year of king Jehoiachin's captivity,
3. The word of the LORD came expressly unto Ezekiel the priest, the son of Buzi, in the land of the Chaldeans by the river Chebar; and the hand of the LORD was there upon him.
4. And I looked, and, behold, a whirlwind came out of the north, a great cloud, and a fire infolding itself, and a brightness *was* about it, and out of the midst thereof as the colour of amber, out of the midst of the fire.

Figure 12.3 Setting sun with halos and features formed by light reflecting and refraction from hexagonal ice crystals in the atmosphere. The sun is surrounded by the 22° halo, flanked on right and left by parhelia (or sundogs). Extending vertically above and below the sun is the sun pillar, and extending horizontally through the sun is the parhelic circle.
Shutterstock

5. Also out of the midst thereof *came* the likeness of four living creatures. And this *was* their appearance; they had the likeness of a man.
6. And every one had four faces, and every one had four wings.
7. And their feet *were* straight feet; and the sole of their feet *was* like the sole of a calf's foot: and they sparkled like the colour of burnished brass.
8. And *they had* the hands of a man under their wings on their four sides; and they four had their faces and their wings.
9. Their wings *were* joined one to another; they turned not when they went; they went every one straight forward.
10. As for the likeness of their faces, they four had the face of a man, and the face of a lion, on the right side: and they four had the face of an ox on the left side; they four also had the face of an eagle.
11. Thus *were* their faces: and their wings *were* stretched upward; two *wings* of every one *were* joined one to another, and two covered their bodies.
12. And they went every one straight forward: whither the spirit was to go, they went; *and* they turned not when they went.
13. As for the likeness of the living creatures, their appearance *was* like burning coals of fire, *and* like the appearance of lamps: it went up and down among the living creatures; and the fire was bright, and out of the fire went forth lightning.

14. And the living creatures ran and returned as the appearance of a flash of lightning.

15. Now as I beheld the living creatures, behold one wheel upon the earth by the living creatures, with his four faces.

16. The appearance of the wheels and their work *was* like unto the colour of a beryl: and they four had one likeness: and their appearance and their work *was* as it were a wheel in the middle of a wheel.

17. When they went, they went upon their four sides: *and* they turned not when they went.

18. As for their rings, they were so high that they were dreadful; and their rings *were* full of eyes round about them four.

19. And when the living creatures went, the wheels went by them: and when the living creatures were lifted up from the earth, the wheels were lifted up.

20. Whithersoever the spirit was to go, they went, thither *was their* spirit to go; and the wheels were lifted up over against them: for the spirit of the living creature *was* in the wheels.

21. When those went, *these* went; and when those stood, *these* stood; and when those were lifted up from the earth, the wheels were lifted up over against them: for the spirit of the living creature *was* in the wheels.

22. And the likeness of the firmament upon the heads of the living creature *was* as the colour of the terrible crystal, stretched forth over their heads above.

23. And under the firmament *were* their wings straight, the one toward the other: every one had two, which covered on this side, and every one had two, which covered on that side, their bodies.

24. And when they went, I heard the noise of their wings, like the noise of great waters, as the voice of the Almighty, the voice of speech, as the noise of an host: when they stood, they let down their wings.

25. And there was a voice from the firmament that *was* over their heads, when they stood, *and* had let down their wings.

26. And above the firmament that *was* over their heads *was* the likeness of a throne, as the appearance of a sapphire stone: and upon the likeness of the throne *was* the likeness as the appearance of a man above upon it.

27. And I saw as the colour of amber, as the appearance of fire round about within it, from the appearance of his loins even upward, and from the appearance of his loins even downward, I saw as it were the appearance of fire, and it had brightness round about.

28. As the appearance of the bow that is in the cloud in the day of rain, so *was* the appearance of the brightness round about. This *was* the appearance of the likeness of the glory of the LORD. And when I saw *it*, I fell upon my face, and I heard a voice of one that spake.

There have been numerous attempts to rationalize this vision, which seems excessively complex, but one of the more interesting of them, and I suspect the most likely, was suggested by Dr. Donald Menzel, director of the Harvard-Smithsonian Observatory in the 1950s. He was also an active UFO debunker from the start of the UFO craze, producing three books and numerous articles on the topic. In the 1950s, he suggested that many reported UFOs were actually unfamiliar optical, meteorological, or astronomical phenomena.[3] As an example, he proposed to account for Ezekiel's vision in the same way—the "wheels" were the 22° and 46° haloes—thus giving "wheels within wheels." The sun pillar and the portion of the parhelic circle provided the horizontal and vertical lines within these haloes that were the spokes of the wheel. The points at which the spokes touched the 22° halo would be made spectacular by the parhelia or sun dogs to the right and left, while the Parry arc and the lower tangent arc would provide extra light at the upper and lower points where the sun pillar intersected the 22° halo.[4] The amber color would not be inappropriate for some of these features, and, if the ice crystals are sufficiently large, the sun dogs and Parry arc can indeed be spectacularly rainbow hued.

As for the animal heads and features, the heads would be the parhelia and other features on the 22° arc, imaginatively compared to animal heads. Menzel claims that it was a common practice to carve the spokes or other features of chariot wheels in the form of animals. I find no support for this claim, but it is not unreasonable.[5] The wheels would be "lifted up" as the sun rose, and they would not turn—the "spokes" would remain horizontal or vertical—as the wheels rose.

Menzel claimed that he had suggested this interpretation to religious leaders and was told it was a plausible inspiration for the vision.[6]

If Menzel sought to head off the pro-UFO forces with this preemptive explanation, the effort backfired. A lot of pro-UFO writers took hold of Menzel's citation of Ezekiel's vision and ignored his naturalistic explanation, using the imagery to bolster their own case for extraterrestrial origins for UFOs. Erich von Däniken was not the first to claim this was the case, but he was one of the more influential. He cited it in his book *Chariots of the Gods*.[7] The idea received an illustrated treatment in the comic book *UFO: Flying Saucers* (1968, Gold Key comics), which sought to present reported cases of UFO visitations and included this purported early instance. Ex-NASA engineer Josef F. Blumrich went even further, writing *The Spaceships of Ezekiel* (Bantam Books, 1974) to provide a possible spaceship design that would conform to the vision. To this day, you can find a great many discussions about whether Ezekiel saw a flying saucer on the Internet. Of course, orthodox religious interpretations reject any of these extraterrestrial explanations.

If the ice crystal halo explanation has merit, there is an interesting wrinkle—the parhelic circle, the sun pillars, and the sun dogs are all effects due to planar ice crystals floating in relatively still air, orienting themselves very nearly horizontally. But the 22° and 46° haloes require *random* orientation of the ice crystals. The upper Parry arc and lower tangential arc require ice needles or columns, rather than plates. How is one to reconcile all of this?

The easiest way is to point out that they have been observed together on numerous occasions. The "St. Petersburg display" of June 18, 1790, was recorded by pharmacist Johann Tobias Lowitz, after whom the Lowitz arcs were named (because of this very report, in fact). Reproductions abound on the Internet. The "Roman display" reported by the Jesuit Christoph Scheiner in 1630 showed many of these effects.[8] There are many photographs and computer illustrations showing the effects on various Internet sites.

By way of explanation, it should be pointed out that it is entirely possible for both types of ice crystal to be present simultaneously, and that crystals need not be located in the same portion of the sky, since the haloes are the result of crystals lying at certain angles relative to the eye and sun, but they are not confined to a particular distance. One can thus get effects from one type of crystal floating nearby and another much farther away. The experimental work of Walter Tapes has shown that crystals responsible for effects can be extremely close by, practically at ground level, or they may be in high-lying cirrus clouds.

Finally, there is another symbol of religious significance that can be attributed to these same phenomena—the Celtic cross.

The Celtic cross looks very much like a standard Christian cross with a circle centered on the intersection. In its earliest form—dating to about the 8th century CE, these were engraved on erected slabs and were attributed to St. Patrick, or to Saint Declan, who, it was thought, combined the Christian cross with the pagan Celtic sun cross as a way to bring the pagans into the Christian fold. No direct evidence supports this claim, and there is a considerable gap between the 5th-century lives of the saints and the first known appearances of the true Celtic cross, but it's easy to see why the attribution was made. The Celtic cross eventually began to be rendered as a solid stone cross with a palpable "halo" around the intersection, and there are many examples in Ireland and Scotland. The motif is common to this day, being frequently used as a grave marker (Figures 12.4, 12.5).

This, of course, raises the question of what the sun cross was that was adapted for Christian use. This was a symbol in use since the Bronze Age throughout Europe, dating back to the second millennium BCE. In its simplest form, it was a simple circle enclosing an equal-armed cross with its intersection at the circle's center, and the radiating arms reaching only to the circle's edge. It's easy to see that this might be the 22°

Figure 12.4 "High cross" or "Celtic cross" from Rock of Cashel, Cashel Ireland. The form of this cross dates back at least to the 9th century CE.
Shutterstock

Figure 12.5 "High cross" or "Celtic cross" from Ireland. The form of this cross dates back at least to the 9th century CE.

Shutterstock

halo, with sun pillar and parhelic circle. But some forms are even more suggestive, with a double circle (which recalls the 46° halo). There are examples with other rays, besides, which could be variations on the basic shape, or might possibly be meant to suggest other ice crystal arcs.[9,10]

Notes

1. https://www.newmexico.org/nmmagazine/articles/post/favorite-sun/
2. https://www.wipo.int/edocs/mdocs/tk/en/wipo_grtkf_ic_38/wipo_grtkf_ic_38_presenta tion_2lorenzo.pdf. See also https://www.ip-watch.org/2018/12/11/indigenous-knowledge-misappropriation-case-zia-sun-symbol-explained-wipo/
3. Dr. Menzel gives a lengthy and detailed list of such phenomena in "UFOs—The Modern Myth," his contribution to *UFO's: A Scientific Debate*, ed. Carl Sagan and Thornton Page (1972), 123–182, specifically 142–143.
4. Actually, Menzel frequently uses what would today be incorrect terminology for many of the effects and arcs observed. I'm not certain how much of this is due to error and how much to a lack of solidification of the terminology when he wrote.
5. The wheels of chariots carved on the walls of the Konark temple in India feature figures carved into the spokes, but the temple dates from the 12th century CE, much later than Ezekiel.
6. Menzel devotes Appendix 7 of "UFOs—The Modern Myth" to this interpretation and devotes the bulk of Chapter 3 of Donald H. Menzel and Ernest H. Taves, *The UFO Enigma: The*

Definitive Explanation of the UFO Phenomenon (New York: Doubleday, 1977) to a fuller discussion of it.

7. Erich von Däniken, *Erinnerungen an die Zukunft: Ungelöste Rätsel der Vergangenheit* (Germany, Econ Verlag, 1968). For an English translation, see *Chariots of the Gods*, trans. Michael Heron (New York: Putnam, 1968).

8. Although the event has been known for a long time, an illustration had been lacking, until one showed up a couple of years ago. See Eva Seidenfaden, "Found: A Diagram of the 1630 Rome Halo Display," *Applied Optics* 50, no. 28 (October 1, 2011): F60–F63.

9. See the examples from Switzerland reported on Wikipedia: http://en.wikipedia.org/wiki/File:Radanhaenger-edited.jpg

10. Zia Pueblo information: https://www.bia.gov/tribal-leaders/pueblo-zia

13
Fiat Lux!

"Let there be light!" These are the first words spoken by God in the book of Genesis. When I first read it in my Child's First Bible, it bothered me that God had created light before the sun, the source of our local light. But the first order of creation by the Judeo-Christian-Muslim God is actually very clever and, at first look, appears unique. And certainly we in the Optics community imagine light as an abstract thing, a collection of photons or of electromagnetic waves, independent of source (or detector) all the time. Besides, if God didn't create light first, how could He see what He was doing?

That's not simple flippancy on my part. The 4th-century St. Basil of Caesarea said that God created light to make the world beautiful, and his contemporary Aurelius Ambrosius wrote: "But the good Author uttered the word 'light' so that He might reveal the world by infusing brightness therein and thus make its aspect beautiful."[1] There is also the much humbler and more practical argument that without light, and God's immediate declaration of light as day and dark as night, there could not be seven days of creation (with the sun and other lights of the sky not created until the fourth day).

Other belief systems and mythologies didn't seem to start with the creation of light. There are stories involving a personification of the sun, often, but not generally of light. In Egyptian myths, for instance, the sun or the sun god arises first from the primeval waters and creates the other deities. In the Babylonian *Enuma Elish*, Apsu and Tiamat, the abyss of fresh and salt waters, are Mother and Father of all else. There are several Greek creation myths, but the one in Hesiod's *Theogeny* is typical, having everything born of chaos, with no light or sun deity appearing for a long time.

Inevitably, through the ages commentators have considered the meaning of the phrase and have suggested that the true significance might be metaphorical rather than literal. It's hard not to interpret "light" as "knowledge," and to interpret the creation of "light" as really the creation of "knowledge." Many institutions of higher learning saw this sort of meaning in the phrase. At least 35 colleges and universities use the phrase "Let There Be Light," "Fiat Lux," or some variation thereof as the mottoes and on their seals. Others, such as Yale, whose motto is "Lux et Veritas," invoke the biblical creation phrase with their choice.[2]

There is also the metaphorical use of light as synonymous with good, as darkness signifies evil, so that in creating light, God could be said to be creating goodness as well. Saint Paul, Thomas Aquinas, and John Bunyan meditated on these meanings. Some saw the dichotomy between light as good and darkness as evil as an equal struggle, and the orthodox Church dubbed this the heresy of Manicheanism. Ambrosius and Basil's poems cited earlier were anti-Manichean works.

Given the primacy and importance of these words—the very first spoken words of the Judeo-Christian-Muslim God recorded are the ones creating light—you would think that these would inspire philosophers and scientists from the very beginning to study light. But this does not seem to have been the case, so far as I can determine. If anything, such associations came long after the work had been done, as in the case of Alexander Pope's 1730 *Epitaph Intended for Sir Isaac Newton*:[3]

> Nature, and Nature's Laws lay hid in Night
> God said *Let Newton Be!* And all was *Light*

"Let There Be Light" or "Lux Fiat" doesn't appear as a title, epigraph, or the stated motivator for any works on light or optics that I am aware of. You would think that someone like the 13th-century Dominican expositor of the rainbow, Theodoric of Freibourg, might have been so inspired.

To return to the original question—is the biblical creation of light as the first thing such a unique thing? Arguably not. In the mythologies of several Native American people in the American Southwest, near the Four Corners area, the people are said to have emerged by climbing upward through a succession of worlds to reach our present one. In many of these one of the first orders of business is to create light to illuminate the dark new world, and so light is created, and a source for the light.[4]

> The First Man and the First Woman decided to make the Fifth World brighter than any of the four lower worlds had been.
> They thought about it for a while. And for a long time they talked about what kind of light they wanted. Until they finally decided to make a sun and a moon.

Sam D. Gill and Irene F. Sullivan observe that "the tale [of the origin of light] is a common motif in Native American creation stories, such as that of the Apache Creation and Emergence. Often the creation of light is attributed to the creator of the world, as in the example of the … Chicksaw…. The Keresans attribute the origin of light to the sons of the female creator Iatiku. The Tsimshian credit Raven."[5]

Notes

1. http://en.wikipedia.org/wiki/Genesis_1:3. See also *A Dictionary of Biblical Tradition in English Literature*, ed. David L. Jeffrey (Grand Rapids, MI: W.B. Eerdmans, 1992), 275–277.
2. For a listing, see here: http://en.wikipedia.org/wiki/Let_there_be_light
3. Although some sources claim this to be his actual epitaph, these lines never were carved in stone for Sir Isaac. His actual epitaph in Westminster Abbey is much longer and less witty, and it is in Latin.
4. Paul G. Zolbrod, *Dine' bahane': The Navaho Creation Story* (Albuquerque: University of New Mexico Press, 1984), 90.
5. Sam D. Gill and Irene F. Sullivan, *Dictionary of Native American Mythology* (New York: Oxford University Press, 1992), 169.

14
Not Worth the Candle

Some praise the sun, and some the moon,
    In eloquence quite grand all;
A fig for both! I'll beat them soon---
The last in May, the first in June---
By that incomparable boon,
    A Spermaceti Candle.

—*Fraser's Magazine*[1]

The candela is the unit of luminous intensity used and defined by the Système International d'Unités. It is, in fact, one of the seven base units, upon which all others are based. As most people in optics are aware, the unit replaces the earlier units of candles and candlepower, being based upon the more systemized and reproducible standard of light having a frequency of 540×10^{12} Hz (corresponding to 555 nm) and has a radiant intensity of 1/583 watt/steradian in that direction. Prior to this, the "new candle" had been defined in terms of blackbody radiation from platinum at its solidification point, but even earlier in terms of the light emitted from a spermaceti candle. Of course, one thinks, the earliest definition of a "candle" derives from an actual candle. What could be simpler or more obvious? It was probably the best easily available source of light they had, easily accessible to anyone needing to reproduce it.

Except that this reasoning is completely wrong. The definition of the spermaceti candle as the standard unit of luminous intensity was adopted in the London Metropolitan Gas Act of 1860, at a time when there were several alternative methods of generating a reliable source of light. The 70 years before that had, in fact, seen a revolution in ways to generate light, especially in the lamps commonly used in the home. Furthermore, even before the Act was passed, spermaceti candles were *already* recognized as a highly variable source of illumination and were not terribly reliable. The spermaceti itself, after all, was the natural product of whales, and it was as likely to vary as much as any other animal product. So why did Britain and, as a result, America choose it as their standard for illumination? Other countries—France, the Netherlands, Germany, Austria, and Scandinavia—used various forms of the item called the Argand lamp as their standard. Why didn't that choice become the international standard?

For households prior to 1800, artificial light usually meant the fireplace or lamps or candles made of animal fat. When I was a child, my grandmother still had such a candle—grayish-brown and smoky. There were candles made of beeswax or bayberry or other clean materials, but these were expensive and considered luxury items. The only other choice was an oil lamp burning vegetable oil.

A major change occurred in 1780 when Swiss inventor Aime Argand developed what came to be called the Argand lamp. It used a tubular wick contained within concentric metal walls. Air fed in through the central space, as well as being drawn up around the outside. A glass chimney helped to shape and maintain the airflow. The resulting flame was larger (which alone could increase the available light), but it also burned at a higher temperature. The lamp could be used with a variety of oils, including whale oil and colza oil (a relative of modern canola oil). Depending upon the oil used, the design, and who did the reporting, Argand lamps could produce between seven and fourteen times the light from a conventional candle. It was a revolution in home lighting. Variations like the Carcel lamp (which used a clockwork pump to deliver heavy colza oil to the wick, without a raised reservoir) and the Buda lamp (which used an oxygen feed in the center, increasing the temperature still higher) represented variations in the basic design.

In 1811, the French chemists Michel Eugene Chevreul and Joseph-Louis Gay-Lussac patented stearin, a substance originally obtained from processing beef fat (although it can be obtained from many different organic oils). Stearin candles were whiter and cleaner than candles made from unprocessed tallow, were harder, and burned with a cleaner flame. Paraffin, discovered in 1832 by Carl Ludwig von Reichenbach, was made an inexpensive commercial product when James Young succeeded in distilling it from coal and oil in 1850. Paraffin was often mixed with stearin to harden it. Such candles became the new home standard.

In addition to these, there was the limelight, invented in the 1820s by English chemist Galsworthy Gurney, and various pre-incandescent forms of electric lighting. These were impractical for home use. But another newly introduced form of lighting was well-adapted to home use. This was the spermaceti candle, made from spermaceti, a waxy substance found in the large head cavity of the sperm whale. In a live sperm whale, the spermaceti is a liquid. It is thought to aid in buoyancy or to act as a sonic lens for the whale's sonar. When the whale has been killed and the body cools, the spermaceti begins to solidify. Sperm whales were prized for this oil, which was emptied by the bucketload from the head of the whale and sold commercially as candles. The candles made from spermaceti were white and hard, and burned with a whiter flame than beeswax or paraffin candles. As with the Argand lamp, the spermaceti candle was endorsed by leading figures of the American Revolution. George Washington experimented with them and determined they were more efficient for lighting his Mount Vernon estate than other types of candles. Benjamin Franklin wrote of their superior quality.

The advantage of spermaceti candles over Argand lamps is that they were simpler and more transportable. Both spermaceti candles and Argand lamps produce brighter, whiter light than candles made of paraffin, beeswax, or stearin. I have verified this by measuring the emission spectrum of candle flames from each type, something I have not seen in the literature. The reasons for the whiter flames of an Argand or spermaceti candle form an interesting study on their own, worthy of an article. Ultimately, it is because the spermaceti itself in one case, and the design of the burner in the other, allow for a higher flame temperature, producing more light at the blue end of the spectrum.

The actual value of the different types of lighting was of only relative interest until public lighting itself became subject to standardization. It was the need to provide a reliable method of measuring public lighting, after all, that impelled Friedrich Richard Ulbricht to invent the integrating sphere.[2] In 1850, the Metropolitan Gas Board of London stated that the emission from a gas flame should equal either the light from 12 wax candles or of 10.3 spermaceti candles. In subsequent rulings, the wax candle was dropped as a standard.[3] Ten years later, in the influential Metropolitan Gas Act of 1860, they specified that it must be a candle, six to the pound, made of spermaceti with enough beeswax added to "break the grain," and burning at 120 grains per hour (7.77 grams).[4]

Why use a spermaceti candle? Certainly the fact that it emitted whiter light made it superior to other candles, but the spermaceti itself, being a natural product made from whale secretions, would seem to make it a poor choice for a standard—one might as well use cow's milk as a standard for whiteness, ignoring the variations of color that result from different breeds, diets, and states of health. Why not use the flame from an Argand lamp of specified fuel, preferably an artificial and standardized fuel. This is what several other European countries did; for instance, France used as a standard a carcel lamp fueled by colza oil.

One reason given for the choice is that even these purified chemical fuels were not, in fact, very pure at the time. With the composition subject to large variations, it was apparently thought that the variation in spermaceti was not significantly different. Another reason is that the output of an Argand lamp depends strongly upon the dimensions of the burner and wick, as well as the glass chimney and the fuel used. Without a great deal of care, the output of an Argand lamp, although notably bright and white, could vary considerably.[5] And wax candles required frequent snuffing and trimming of the wick to produce the maximum illumination, something that spermaceti candles did not require.[6]

However, a spermaceti candle produced a flame of uniform size, and the wick needed no trimming. Apparently it was felt that a spermaceti candle flame achieved reproducibility with a minimum of engineering fuss.

It didn't work out that way in practice, however. Even before the standard was adopted, some complained about the nonreproducibility of the "standard." In fact, William Crookes (inventor of the Crookes Tube and the Crookes radiometer[7]) complained about the use of the spermaceti candle as a standard. "Assuming that the true paliamentary candle is obtained, made from the proper materials and burning at the specified rate, its illuminating power will be found to vary with the temperature of the place where it has been kept, the time which has elapsed since it was made, and the temperature of the room wherein the experiment is tried. . . . It would be impossible for an observer on the continent, ten or twenty years hence, from a written description of the sperm candle now employed, to make a standard which would bring his photometric results into relation with those obtained here."[8] "It is impossible to make a candle which will unfailingly burn at 120 grains an hour, or within five per cent of it," complained an 1898 report.[9] "The average rate at which a sperm candle burns per hour is 132 grains," noted the *Journal for the Society of the Arts*. "The number 120, which is fixed as the standard for comparison in all Acts of Parliament relating to the subject is, therefore, too low, for it is rarely if ever obtained in practice."[10] "The standard sperm candle has long been doomed as a unit of light," declared *The Photographic News* in

1882. "Anyone who has given due attention to a burning candle knows how variable are its conditions of consumption." Moreover, an even rate of consumption does not guarantee standard illumination. "A standard candle should burn away at a rate of 120 grains per hour; but, as everyone knows, a certain consumption of material does not mean a certain light in candle-burning."[11] In practice, the candle used was split in two at the center and the candle material trimmed back from the wick. The rate of burning was determined, and a proportional correction made. If the rate fell below 114 grains per hour, or above 125, the test was to be rejected. The candles were allowed to burn for 10 minutes before being used in tests.[12,13]

The candle thus defined was ratified as an international standard at the International Electrotechnical Conference in Paris in 1881.[14] Nevertheless, most countries continued to use their own references. Even in Britain, the candle as a standard did not last long. A. Vernon Harcourt, the senior Metropolitan Gas referee, looked for a stabler reference and used the burning of air saturated with pentane vapor. The flame produced by this was metered and found to produce a flame equal to that of a spermaceti candle. In 1898, the Metropolitan Gas Commission accepted a version producing the illumination equal to 10 spermaceti candles for statutory tests, although the standard spermaceti candle continued to be used for most testing.[15]

Germany changed to the use of a paraffin standard candle in 1869 (with its manufacture standardized in 1871).[16] In 1885, Germany use the Hefner lamp as a standard. Fueled by amyl acetate, which is so volatile that it evaporated before the wick could char, it was felt to provide a reliable standard. It was adopted in other European countries as well, but the candle continued to be the reference in the United States and England.[17]

The Dutch were not satisfied with the stability of the Hefner lamp, and in 1894 adopted a variation of Harcourt's lamp, fueled with a mixture of ether and benzol.[18]

The variation of candle output was finally demonstrated by C. H. Sharp and C. P. Matthews using a bolometer and a recording galvanometer in 1896. They measured the variability in the output of a standard British candle and the German paraffin candle, and compared them with that of a Harcourt pentane lamp, much to the detriment of the candles.[19]

In 1946, the Commision Internationale d'Éclairage proposed a new international standard for illumination to provide an acceptable international standard. The new candle was established with the definition that a blackbody radiating at 2041.4K (the solidification temperature of platinum) would be equal to 60 new candles per square centimeter. The definition was accepted in 1948 at the ninth meeting of the Conférence générale des poids et mesures (CGPM) at Sèvres, at which time the name *candela* (Latin for "candle") was adopted as the name for the new unit, although the term "New Candle" continued to be used until 1967, when the CGPM revised the definition to specify that it was the luminous intensity perpendicular to a surface of a blackbody 1/600,000 of a square meter in area at the temperature of freezing platinum and a pressure of 101,325 newtons per square meter.[20]

In 1979, the definition was redefined as

the luminous intensity, in a given direction, of a source that emits monochromatic radiation of frequency 540×10^{12} hertz and that has a radiant intensity in that direction of $1/_{683}$ watt per steradian.

So, in brief, the candle was adopted as standard, despite its problems with standardization and its apparent retrograde orientation, because it was seen as no worse than other sources, and because its output was more reliably reproducible than a wax candle whose output depended upon the state of trimming of its wick, or an Argand-type lamp dependent upon the size and setting of its wick. A spermaceti candle was easily obtained and would provide uniform light without much care for the state of its burning.

Notes

1. Anonymous poem, *Fraser's Magazine* 6, no. 36 (December 1832): 686–688.
2. Stephen R. Wilk, "Ulbricht's Kugelphotometer," *Optics & Photonic News* 25, no. 1 (January 2014): 26–28; and Chapter 5 in this volume.
3. Chris Olter, *The Victorian Eye: A Political History of Light and Vision in Britain, 1800–1910* (Chicago: University of Chicago Press, 2008), 162.
4. Another source says that it was typical in manufacturing spermaceti candles to add 3% wax "to prevent crystallization." See Charles Wentworth Dilke, *Exhibition of the Works of Industry of All Nations, 1851: Reports of the Juries* (London: William Clowes and Sons, 1852), 627. https://books.google.com/books?id=gk3-bG2_tXUC&pg=PA627&dq=%22 spermaceti+candle%22&hl=en&sa=X&ved=0CD8Q6AEwA2oVChMIzduH5t6axwIVS Y0NCh2Uuwri#v=onepage&q=%22spermaceti%20candle%22&f=false. See also George Gore, J. Scoffern, and Marcus Sparling, *The Chemistry of Artificial Light, Including a History of Wax, Tallow, and Sperm Candles* (London: Houlston and Stoneman, 1856), 43.
5. See John Mason Good, Olinthus Gregory, and Newton Bosworth, *Pantologia: A New Cyclopedia* (London, 1813), entry on "Lamp": "Count Rumford, as we have seen, used the Argand lamp as a standard for comparison; but as the intensity of its light varies according to the height of the wick, Mr. Hassenfritz preferred a wax candle, making use of it soon after it was lighted" [and presumably before the wick required trimming]. https://books.google. com/books?id=aeZTAAAAYAAJ&pg=PT581&dq=%22argand+lamp%22&hl=en&sa=X &ved=0CEgQ6AEwB2oVChMI18KbsuDJxwIVz36SCh20ygTX#v=onepage&q=%22arg and%20lamp%22&f=false
6. See, for instance, "Proceedings of the Second Annual Meeting of the American Gas Institute held at the New Willard Hotel, Washington, DC, October 16, 17, and 18, 1907," *The Progressive Age*, January 1, 1908, 12–27, especially 21. Appendix D: "Notes on English Standard Candles" can be found at https://books. google.com/books?id=XipKAQAAMAAJ&pg=PA21&dq=Metrolpitan+Gas+Act+1860+sperm aceti&hl=en&sa=X&ved=0CE8Q6AEwCWoVChMImObc4YPexwIVgk2ICh3NQA3b#v=onepa ge&q=Metrolpitan%20Gas%20Act%201860%20spermaceti&f=false. It is also available, more legibly, at *Proceedings of the American Gas Institute*, Second Annual Meeting, published in 1908 at https:// books.google.com/books?id=IMQSAAAAYAAJ&pg=PA499&dq=Metrolpitan+Gas+Act+186 0+spermaceti&hl=en&sa=X&ved=0CBwQ6AEwADgKahUKEwj93ITimt7HAhWWK4gKHV- pACY#v=onepage&q=Metrolpitan%20Gas%20Act%201860%20spermaceti&f= false
7. Stephen R. Wilk, "Crookes' Radiometer," *Optics & Photonics News* 18, no. 9 (September 19, 2007), 17; and Chapter 17 in *How the Ray Gun Got Its Zap!* (Oxford University Press, New York 2013).
8. William H. Crookes, "On the Measurement of Luminous Intensity of Light," *The Chemical News and Journal of Physical Science* 18, no. 449 (September 1868) : 123– 128. https://books.google.com/books?id=RQBLAAAAYAAJ&pg=PA124&dq=%22sp

erm+candle%22&hl=en&sa=X&ved=0CB0Q6AEwAGoVChMIwbea9_WEyQIViHo-Ch2PUQwp#v=onepage&q=%22sperm%20candle%22&f=false   Also here, July 17, 1868: https://books.google.com/books?id=9jNCAQAAMAAJ&pg=RA1-PA26&dq=%22sperm+candle%22&hl=en&sa=X&ved=0CCIQ6AEwAWoVChMIwbea9_WEyQIViHo-Ch2PUQwp#v=onepage&q=%22sperm%20candle%22&f=false

9. Cited in *The Progressive Age*, 21 or in *Proceedings of the American Gas Institute*, 502.

10. "Quality of the City Gas," *Journal of the Society of Arts* 8, no. 380 (March 2, 1860): 256–257. https://books.google.com/books?id=4TI9AQAAIAAJ&pg=PA256&dq=%22sperm+candle%22&hl=en&sa=X&ved=0CBsQ6AEwADgKahUKEwi04qGU9ITJAhXMdj4KHSQEBng#v=onepage&q=%22sperm%20candle%22&f=false

11. "A Unit of Light," *The Photographic News: A Weekly Record of the Progress of Photography* 26 (June 2, 1882): 307–308.

12. From Georg Lunge (ed.), *Technical Methods of Chemical Analysis*, trans. Charles Keane, ed. Van Nostrand , vol. II, part II (1911): 698–699.https://books.google.com/books?id=15mcWFUkxRgC&pg=PA698&dq=Metrolpolitan+Gas+Act+1860+spermaceti&hl=en&sa=X&ved=0CC4Q6AEwA2oVChMImObc4YPexwIVgk2ICh3NQA3b#v=onepage&q=Metrolpolitan%20Gas%20Act%201860%20spermaceti&f=false; see also W. J. Dibdin, "Public Lighting by Gas and Electricity," Chapter 1, "Artificial Light, Its Source and Measurement," *Sanitary Record and Journal of Sanitary and Municipal Engineering* 24 (Nov. 17, 1899): 429–430.   https://books.google.com/books?id=T3lIAAAAYAAJ&pg=PA429&dq=Metrolpolitan+Gas+Act+1860+spermaceti&hl=en&sa=X&ved=0CDoQ6AEwBWoVChMImObc4YPexwIVgk2ICh3NQA3b#v=onepage&q=Metrolpolitan%20Gas%20Act%201860%20spermaceti&f=false

13. On burning rates, John Thomas Cooper in 1838 claimed that wax candles would burn at a rate of 122 grains per hour, spermaceti at 129, and stearin at 156. *The Lancet* 2 (July 28, 1838): 613–614. https://books.google.com/books?id=OBVAAAAAcAAJ&pg=PA614&dq=%22spermaceti+candle%22&hl=en&sa=X&ved=0CFgQ6AEwBzgKahUKEwi6_oGY4ZrHAhUE7IAKHfuaDNY#v=onepage&q=%22spermaceti%20candle%22&f=false Andrew Ure, in 1839, claimed that a spermaceti candle would burn at 132 grains an hour. See "Experimental Researches upon the Relative Illuminating Powers of Different Lamps and Candles, and the Cost of the Light Afforded by Them," *The Civil Engineer and Architect's Journal and Scientific and Railway Gazette* II (September 1839): 328–331. His candles were three to the pound and 9/10" in diameter. https://books.google.com/books?id=KwsAAAAAMAAJ&pg=PA330&dq=%22spermaceti+candle%22&hl=en&sa=X&ved=0CC8Q6AEwADgUahUKEwis8-3A4prHAhUE8IAKHc78DcE#v=onepage&q=%22spermaceti%20candle%22&f=false

14. G. Jerrard and D. B. McNeill, *A Dictionary of Scientific Units*, 5th ed. (New York and London: Chapman and Hall, 1986), 24.

15. Lunge, *Technical Methods of Chemical Analysis*, 698 and 699–701.

16. See *Proceedings of the Annual Meeting* in note 6.

17. This reference claims that one reason for its selection was the difficulty in obtaining pure pentane for a Harcourt lamp, and that the amyl acetate was purer and more consistent than spermaceti or colza oil. Wilbur M. Stine, "The Candle Power of Arc and Incandescent Lamps: The Amyl Acetate Lamp," *American Electrician* 11, no. 5 (May 1899). https://books.google.com/books?id=zQU0AQAAMAAJ&pg=PA209&dq=%22amyl+acetate%22&hl=en&sa=X&ei=iWyHVcDrNZPcgwTw6IKADQ&ved=0CEQQ6AEwBw#v=onepage&q=%22amyl%20acetate%22&f=false

18. Groves and Thorp's *Chemical Technology —or—Chemistry Applied to Arts and Manufactures. Vol. IV—Electric Lighting and Photometry* by A. G. Cooke (Electric Lighting) and W. J. Dibdin (Photometry) (Philadelphia: P. Blakiston's Son and Co., 1903), 335–336.

19. "A Report on Standards of Light Presented to the American Institute of Electrical Engineers," *Trans A.I.E.E.* 8 (1896), reported in Edward L. Nichols, "Some Notes on the Early History of Standards of Light," *Transactions of the Illumination Engineering Society* 5, no. 9 (December 1910): 842–853.

20. The radiation from platinum at its freezing point was first suggested as a standard by J. W. Draper in *American Journal of Science* 2, no. 4 (1847). It was suggested several times until it was adopted in 1884 as a fundamental unit of luminosity, but it was not accepted as a standard until much later. See Edward L. Nichols, "Some Notes on the Early History of Standards of Light," *Transactions of the Illuminating Engineering Society* 5, no. 9 (December 1910): 842–853, esp. 850–852. https://books.google.com/books?id=4S06AQAAMAAJ&pg=PA844&lpg=PA844&dq=%22Metropolitan+Gas+Act%22+1860+spermaceti&source=bl&ots=2zSFN3WWQo&sig=oHabQK4OrpH2Vvb_yQ8DD4gjcYg&hl=en&sa=X&ei=X8NxVYOxG8j9oAT3xoDQDg&ved=0CDsQ6AEwBg#v=onepage&q=%22Metropolitan%20Gas%20Act%22%201860%20spermaceti&f=false

15
Why Are Candle Flames Yellow?

Why, indeed, are candle flames yellow? This is one of those questions that I thought I knew the answer to, but it turns out that I wasn't correct—there are nuances in the answer.

The question arose when I was researching the spermaceti candle. Numerous sources stated that the light from a spermaceti candle was brighter and whiter than that from other types of candles. A similar claim was repeatedly made for light from an Argand lamp, in which liquid fuel is burned in a lamp with a cylindrical wick, which draws air through the center as well as around the outside. Both types of candles were held to give from seven to fourteen times as much light as an ordinary candle.

Yet how could this be? In all cases—tallow candles, beeswax candles, spermaceti candles, Argand lamps fueled by vegetable oil—we are simply burning hydrocarbons. Surely the temperatures of the flames are similar. Furthermore, as readers familiar with Michael Faraday's classic lecture *The Chemical History of a Candle* know, the yellow part of the candle flame is due to the glow of common soot heated by the flame. Carbon black is almost a perfect blackbody in the visible, so surely the glow we are seeing is the blackbody radiation from particles heated to the temperature of the flame. [1]

It turns out that there are several mistakes and misconceptions in this explanation. I didn't realize this until I used a spectrometer to measure the spectrum of a flame myself, because I couldn't find any such measurements in the literature. I finally obtained a spectrum of a homemade spermaceti candle and of an Argand-type kerosene lamp. But when I tried to fit them to a blackbody curve of what was supposed to be the typical temperature of a candle flame, I found that it was not at all similar. Even worse, I could not make it appear to be similar by varying the temperature. The shape of the emission was altogether different. It did, however, confirm that the emission spectra of the spermaceti candle and the kerosene lamp extended further into the visible, producing a whiter flame. The Argand lamp may have a larger flame, but it puts out more light in the visible as well.

The reason that the shape of the spectrum differs from that of a blackbody is that, while soot is indeed an ideal blackbody radiator *in bulk*, the soot in a candle flame is of extremely small size, mostly smaller than the wavelength of light. Its emissivity thus varies with an inverse power of the wavelength. What this means is that a candle flame has much more of its output shifted to shorter wavelengths than a blackbody made of, say, bulk soot at the same temperature. [2]

The implications of this for civilization are important. The peak emission for both a blackbody at 1400°C or a candle flame at 1400°C (1673 K) are both in the infrared, but the blackbody peaks at 1.73 microns, and its appearance is a dull red glow, represented most nearly by 595 nm, while the candle flame at the same temperature appears to be a yellow glow best represented by light at about 580 nm. This would require a blackbody temperature of about 3500K, and has more of its emission spectrum in the visible, giving it the familiar yellow glow that has lit the world after the sun has gone down since Neolithic times. If we had had to rely on blackbody light from a 1673K blackbody, civilization would not have advanced as it has—there would have been much less available light at night from campfires, oil lamps, and candles. Unless some early genius had discovered an alternative (such as using a heated element dipped into sodium salts to provide the bright sodium D lines), our progress would probably have been much slower.

As for the question of how the spermaceti candle and the Argand lamp can have a whiter spectrum, there are two possible causes—the size distribution of soot particles could be different, with more of the smaller particles (thus pushing the emission to even shorter wavelengths), or else the flames might burn hotter, thus shifting the peak to shorter wavelengths. (It is possible that both mechanisms may, in fact, be present).[3,4]

The superior light output of the Argand lamp has long been explained as due to the fact that the air being drawn both through the center of the wick as well as around the outside (and being regulated by the shape of the glass chimney of the lamp) produces a better-fed and hotter flame. This does appear to be the correct explanation, and it was Aimee Argand's own reasoning in creating the lamp. According to Professor Peter Sunderland of the University of Maryland, soot particles tend to be the same size, regardless of source—about 20–40 nm. The largest difference between flames is the temperatures, which vary from region to region of the flames, and can be influenced by the construction of the lamp and by the source of fuel. The spermaceti burns at a higher temperature, but an Argand lamp allows more pedestrian fuels to achieve higher temperatures as well. And the greater flame area, while not the main contributor to the brightness, is still a factor in the overall light produced.[5,6]

Many thanks to Prof. Peter Sunderland of the University of Maryland, Marcus Chaos of FM Global, and many others that I pestered with questions. None of my errors in this article should be attributed to them.

Notes

1. On the emissivity of fine soot, see V. Robert Stull and Gilbert N. Plass, "Emissivity of Dispersed Carbon Particles," *Journal of the Optical Society of America* 50, no. 2 (1960): 121–129; Sakae Yagi and Hiroshi Iino, "Radiation from Soot Particles in Luminous Flames," *Eighth International Symposium on Combustion* 8, no. 1 (1961): 288–293; Xie Huanein, "A Study of Mechanism of Radiation in Luminous Flames," in *Fire Safety Science*, ed. Peter Mullinger and Barrie Jenkins, *Industrial and Process Furnaces*, 2nd ed. (Elsevier, 2014), 461–464. http://www.iafss.org/publications/aofst/1/461/view/aofst_1-461.pdf. A good reference is C. L. Tien and S. C. Lee, "Flame Radiation," *Progress in Energy and Combustion Science* 8, no. 1 (1982): 41–59.
2. On the size of soot particles, see William C. Hinds, *Aerosol Technology: Properties, Behavior, and Measurement of Airborne Particle*, 2nd ed. (Wiley Interscience, 1982, 2012); see also Sheldon K. Friedlander, *Smoke, Dust, and Haze: Fundamentals of Aerosol Dynamics*, 2nd ed. (Oxford University Press 2000).

3. The only published record of the emission spectrum of an Argand burner I have seen was sent to me by Marcos Chaos of FM Global, and it is from "The Spectrum of an Argand Burner" by S. P. Langley. It appeared in *Science* 1, no. 17 (June 1, 1883): 481–484. The fuel was "house gas" (presumably coal gas), and the spectrum depicted peaks at 1.5 microns.

4. The original speculation on incandescent soot as the source of the yellow glow in candle flame is Sir Humphry Davy, "Some New Researches on Flame," *The Philosophical Magazine and Journal* (series 1) 50, no. 231 (July 1817): 3–26.

5. On a history of lighting, see Davis DiLaura, "A Brief History of Lighting," *Optics & Photonics News* 19, no. 90 (September 2008): 23–28.

6. Michael Faraday's *A Chemical History of a Candle* has been published several times since he first delivered the lectures at Christmas in 1859–1860. It first appeared in 1861, and it remains in print after all this time. It can also be found on the Internet, including on Project Gutenberg at http://www.gutenberg.org/ebooks/14474. There are also various interpretations of his lectures on YouTube. Jearl D. Walker's more recent "The Physics and Chemistry Underlying the Infinite Charm of a Candle Flame" appeared in his Amateur Scientist column in *Scientific American* 238, no. 4 (April 1978): 154–162.

PART II
WEIRD SCIENCE

16
Brown

Ernst von Brücke was a physician, a physiologist, and a physicist, a cofounder of the
Physikalische Gesellschaft in Berlin, a noted optical physicist, and an influence on both
Hermann von Helmholtz and Sigmund Freud.[1] That he could not account for the
position of brown relative to the known colors, then, shows the state of the understanding of color in the middle of the 19th century.

The question he asked is a not uncommon one, judging from how often a quick
Internet search turns it up online. If you look at a spectrum, or a CIE chromaticity
diagram, there is no brown. "All the colors of the rainbow" is a phrase that's supposed
to mean all possible colors. But if the rainbow doesn't include brown, then what is it?

Brown is the color of dirt, the color of untidiness. The Impressionists in the late
19th century, drunk on the new bright colors released by chemistry, decried the
"brown gravy" that dominated Renaissance art and used the new theories and chemistry of color to repaint the world in bright colors. Who wanted dull, ordinary brown?
A survey conducted by the firm GfK Bluemoon of Australian smokers determined
that a brown shade represented by the Pantone number 448C is aesthetically "the
most unappealing." Australian law mandates that cigarette packaging be changed
from bright and attractive colors to unappealing ones, and Pantone 448C appears to
be the winner (or loser). It is, of course, a dark, muddy brown. (Its sRGB equivalent is
(74, 65, 42).) Respondents associated the color with "dirty," "tar," and "death." One researcher posited that its resemblance to that of human waste is responsible for its low
rating. In a similar case, the American color consultant Faber Birren suggested to a
department store in the 1960s that they could shorten employees' bathroom breaks by
painting the walls of the restrooms in a similar color. Even 448C has its supporters, but
clearly a lot of people disapprove of it. Brown gets no respect.

When I was a kid, I soon learned that I could make a serviceable brown by mixing
the colors yellow and purple, or by mixing red and green. Neither combination was
ever represented on those color wheel charts we saw in school. The spectrum of
brown filters from companies like Roscolux show multiple peaks, [2]notably in the violet, blue, green, yellow-orange, and deep red. Albert Henry Munsell, developer of
the color system that bears his name, defined brown as "A Dark color, inclining to
Red or Yellow, obtained by mixing Red, Black, and Yellow."[3] In the sRGB system, one

definition of "brown" is (R,G,B) = (150, 75, 0), which is a 2:1 mixture of red and green, much as I discovered in my youth, albeit with additive rather than subtractive colors.

From all this, it is evident that brown is really a sort of dark yellow or dark orange. It's not really absent from the spectrum, but it's not obviously represented because most spectral representations—the rainbow, color wheels, the CIE Color Diagram— are essentially two-dimensional representations, and lack the third dimension of brightness (or, in Munsell's terminology, *value*) to show how dark the color is. This is true even of light browns.

Since von Brücke's time, the color system Munsell developed in the first quarter of the 20th century, the CIEXYZ, CIELAB, and other systems based on the work of the Commission Internationale de l'éclairage, and the various RGB systems used by color displays, have given us a technical vocabulary for expressing the three degrees of freedom for color that includes, in addition to the familiar spectrum, the more complex colors of purples, browns, grays, and flesh tones.

So now we have a handle on brown, its spectrum, and how it can be represented. But the first real customer for the means of quantitatively describing browns was a surprising one—the United States Department of Agriculture. In 1894, the USDA established a Division of Agricultural Soil within, of all places, the Weather Department of the USDA. The Agricultural Appropriations Act of 1896 gave the USDA authority to examine and publish results about soil, which had become a topic of great interest in the 19th century. Work by American, German, and Russian scientists on the origins, nature, and management of soil provided information on how to produce good crops. This was seen as essential to the USDA mission of advising farmers. In 1899, what would become the National Cooperative Soil Survey began with surveys in four geographic regions under Milton Whitney, the first chief of the Division of Agricultural Soil.[4]

Soil experts described the soils they observed in their reports, but it was clear that there was little agreement between designations, or consistency in the naming. Even after the USDA issued a report in 1925 with a set of 20 standard names, and the issuing of the first Soil Survey Manual in September 1937, with its listing of standards, the agreement was poor. Soil scientists visiting Washington or attending the 1939 New Orleans meeting of the Soil Science Society of America were asked to provide names for 250 standard samples of soil the USDA had gathered. Each sample received, on average, 12 different designations. Participants who tried to name samples more than once "were unable completely to duplicate names."[5]

It was to provide a more consistent and reproducible method of naming soil colors that the USDA adopted what was essentially the Munsell system, using a recommendation by the National Bureau of Standards and the Inter-Society Color Council. At the time it was first proposed that they do this, there really was no alternative scheme. Later, the USDA derived their own charts and formulas for interpolation between shades, and for converting to the CIE color notation. The USDA made up booklets of pages with color samples on them that could be compared directly with soil samples "in the field." Says the "Preliminary Color Standards" report of 1941, "Few soil colors will match perfectly the colors on the charts, but all colors should be easily located as nearer to one color than to another."[6]

Munsell books of soil colors and color charts are still printed and sold, with color chips having cutouts through which the soil can be observed and compared. The books are used not only by soil scientists and agricultural experts, but by geologists, archaeologists, and other experts trying to characterize soils. Brown, as the color of soil, is getting the respect and treatment it deserves.

Notes

1. Quoted from Ernst W. R. von Brücke, "On the Existence of the Color Brown," *The American Journal of Science and Arts* 7, no. 129 (May 1849), which, in turn, cites *Philosophical Transactions of the Royal Society* (which is responsible for the translation) 33, no. 281 (1848), which cites the original publication in the *Poggendorff Annalem der Physik* 7 (1848). https://books.google.com/books?id=zBzZYe9n--EC&pg=PA129&dq=%22color+brown%22&hl=en&sa=X&ved=0ahUKEwiyoqC1ppHNAhVDWz4KHSIRACQ4ChDoAQgqMAM#v=onepage&q=%22color%20brown%22&f=false. On von Brücke himself, see https://en.wikipedia.org/wiki/Ernst_Wilhelm_von_Br%C3%BCcke

2. Accessed June 10, 2016, http://www.cnn.com/2016/06/09/health/pantone-448c-color-cigarette-advertising/index.html. The CNN piece cites Angela Wright, a color consultant and author of *The Beginner's Guide to Colour Psychology* (London: Kylie Cathy Limited, 1995).

3. For examples, Roscolux #3405, Roscolux #3406, or Roscolux #99 "Chocolate"; see https://www.rosco.com/filters/roscolux.cfm, which gives a rough idea of the color within the sRGB system and includes links to the spectra of the filters. It also includes CIE plots which show the colors of these as squarely within the region we normally call "orange."

4. Albert H. Munsell, *A Color Notation*, 2nd ed. (Boston: George H. Ellis Co., 1926), Glossary.

5. See *Soil Survey Manual* published by the United States Department of Agriculture, http://r.search.yahoo.com/_ylt=A0LEV0ug3FRX.IsA1SlXNyoA;_ylu=X3oDMTByNXM5bzY5BGNvbG8DYmYxBHBvcwMzBHZ0aWQDBHNlYwNzcg--/RV=2/RE=1465208097/RO=10/RU=http%3a%2f%2fwww.nrcs.usda.gov%2fInternet%2fFSE_DOCUMENTS%2fnrcs142p2_050993.pdf/RK=0/RS=a3VX0C1cAs0uPz1LjG_vUzqyO8I-. See also https://en.wikipedia.org/wiki/National_Cooperative_Soil_Survey and "Preliminary Color Standards" in the next note.

6. T. D. Rice, Dorothy Nickerson, A. M. O'Neal, and James Thorpe, *Preliminary Color Standards and Color Names for Soil Samples*. USDA Miscellaneous Publication #425 (September 1941). https://books.google.com/books?id=Mzj2dClpcZQC&pg=PA8&dq=%22munsell%22&hl=en&sa=X&ved=0ahUKEwi6pKjZqZHNAhVMET4KHUHIBDc4FBDoAQg3MAY#v=onepage&q=%22munsell%22&f=false

17

Indiana Jones and the Temple of Light

Sometimes an academic joke or deliberately over-the-top speculation can get you in trouble. People not in on the joke or who are unaware of how outrageous the idea is can take it literally, and then it can take on a pseudoscientific life of its own.

Consider, for example, Sir Joseph Norman Lockyer (1836–1920), author of *The Dawn of Astronomy* (1894), in which he argues that many temples in the ancient world were aligned along critical sunrise directions.[1] He observed on p. 180 that

> in all freshly-opened tombs there are no traces whatever of any kind of combustion having taken place, even in the innermost recesses. So strikingly evident is this that my friend M. Bouriant, while we were discussing this matter at Thebes, laughingly suggested the possibility that the electric light was known to the ancient Egyptians.

Of course, eventually people *did* take it seriously. A crowd of "anomalous facts" writers in the 1950s collected such tidbits from older writers, and then in 1970 Erich van Daniken burst forth into bestsellerdom by claiming that ancient astronauts had made a gift of advanced technology to ancient people. This was just one of his examples, bolstered by images such as the supposed "Egyptian light bulb" from Denderah and the "Baghdad battery."

Lockyer didn't really think that electric lights were responsible, of course. He took his cue from an earlier speculator on archaeoastronomy, Charles François Dupuis (1742–1809), who wrote in his 1795 book *Origine des tous les Cultes ou Religion Universelle* (Vol. 1, p. 459) that a temple at Heliopolis was "flooded all day long with sunlight by means of a mirror."[2] This was undoubtedly gleaned from some ancient record, since all temples at Heliopolis in Egypt were scavenged and destroyed centuries before Dupuis wrote, but he did not identify his source. But Lockyer agreed with him, saying, as we have quoted, that there were no signs of combustion. "Doubtless all the inscriptions in the deepest tombs were made by means of reflected sunlight," wrote Lockyer. "With a system of fixed mirrors inside the galleries, whatever their length, and a movable mirror outside to follow the course of an Egyptian sun and reflect its beams inside, it would be possible to keep up a constant illumination in any part of the galleries, however remote."

Lockyer's word on this carried considerable weight, because he wasn't simply someone speculating about Egyptian archaeoastronomy. He was the first Professor of Astronomical Physics at the Royal College of Science at South Kensington. He was the founder and the first editor of the journal *Nature*. For good measure, he was codiscoverer of the element Helium. Lockyer had impressive cross-disciplinary scientific chops.

Nevertheless, modern optical scientists and Egyptologists aren't impressed with his theories about tomb lighting. There is plenty of soot in many underground chambers, and lighting materials have been found, along with representations of them. In addition, a well-made and kept flame need not be extremely sooty. On the other side, as we discussed in connection with Archimedes's mirror,[3] copper or bronze mirrors are not extremely good reflectors, and mirrors that are not absolutely flat (or only very slightly curved) will rapidly lose much of your light. Even a perfectly flat mirror must cope with the fact that the finite angular extent of the sun means that the rays will diverge.

In brief, it seems to me that you could use a single mirror or perhaps a couple of them to direct light a short distance into a deep chamber, but it's not very satisfactory, and not trivial to do. It doesn't seem like a likely thing for the Egyptians to have done.

Nevertheless, Lockyer had repeated the suggestion, which got repeated by others citing him, and eventually found its way into pop culture. The 1954 movie *Secret of the Incas* features Charlton Heston as adventurer Harry Steele looking for lost treasure whose location is revealed by sunlight reflecting from mirrors placed outside and inside an Incan tomb. The movie made a strong impact on Steven Spielberg. His hero Indiana Jones dressed almost exactly like Heston's Harry Steele, sporting a leather jacket, fedora, tan pants, shoulder bag, and three-day stubble. The "Well of Souls" sequence in *Raiders of the Lost Ark* (1981), where the treasure location is revealed by a beam of sunlight, seems to be a direct result.

But the really spectacular pop culture result came in the 1999 film *The Mummy*, where Brendan Fraser's Rick O'Connell (who dresses pretty similarly to Harry Steele and Indiana Jones) enters a tomb and fires his gun at the mounting of a mirror, sending it into an alignment that directed sunlight into the chamber and didn't merely light a spot, but illuminated the entire room. It was Dupuis's fantasy brought to cinematic life.

After a string of blockbusters like that, the idea started showing up everywhere—in comic books, literature, more movies, video games, and the like. It has achieved its own page on the website *TV Tropes*, a sure sign that the concept has "made it" in the world of pop culture.[4] The cable TV show *Mythbusters* even tried to duplicate the effect, from its incarnation in *The Mummy*. Their ruling: the idea of illuminating a tomb with mirrors was "plausible but ridiculous." And the lighting level was extremely low.[5]

Granted that the idea is far-fetched as a way of lighting Egyptian tombs, however, we should take it a step farther and ask, "Did anyone else later in history do this?" After all, mirror technology improved through the years. The Romans had developed metal-backed glass mirrors large enough to display a full human figure. By 1679, they could line the walls of the Hall of Mirrors at the Palais de Versailles with them. Did anyone else use mirrors to direct sunlight into buildings?

The only such case I have found before modern times is the suggestion that the *Natatio* at the Roman Baths of Caracalla, constructed at the beginning of the third century CE, had bronze mirrors hung in place of a roof so that they could direct sunlight down onto the bathers. The notion is ubiquitous on the Internet, especially on Roman tour sites.

Could this be true? I noticed that almost none of the Internet sites that wrote about this remarkable phenomenon gave citations or references to back it up. It appeared that they were simply citing each other, which is how Internet rumors propagate. I have written about a similar case recently with William Byler and the invention of the black light.[6] I contacted Janet Delaine, Emeritus Fellow of Classics at Oxford, and the author of the definitive work on the archaeology and structure of the Baths of Caracalla, who was very surprised to learn of this recent development. "There is absolutely no truth in it at all," she wrote.

So where *did* the idea come from? One of the few Internet sites to give a reference—the Wikipedia article on the Baths of Caracalla[7] —cited Leland M. Roth's book *Understanding Architecture*.[8] Roth's book did, indeed, mention the possibility but gave no reference or citation either. Checking to see that Professor Roth was still around, I wrote to him. He graciously responded, saying that he drew the information from Paul MacKendrick's book *The Mute Stones Speak: The Story of Archaeology in Italy*.[9] MacKendrick does not directly identify the source but does have a set of references. He was guarded in his use of the idea, noting that "The evidence for the bronze plates and the sun room is not archaeological but literary, and, chiefly because the literary source has little or no idea of what he was talking about, has raised apparently insoluble controversy."

The relevant citation that MacKendrick drew on was evidently Hugh Plommer's *Ancient and Classical Architecture*.[10] Citing the work of "Miss Broedner," he says that she maintained, in her analysis of the Baths, that the *tepidarium* and the *calidarium* were "entirely covered by a light, suspended flat roof of bronze plates designed to catch the sunlight for the benefit of the athletes below." It's clear, however, that Plommer isn't persuaded by Broedner, since, after arguing against some of her interpretations, he says "One must leave the detailed refutation of Miss Broedner to an investigator patient enough to endure the opera."

"Miss Broedner" is actually Erika Brödner, a scholar whose work *Untersuchungen an den Caracallathermen*[11] speculates upon the setup of the Baths of Caracalla. The information about the possible bronze plates derives from a brief notice of the baths written in the early 3rd century CE work *Historia Augusta*.[12]

> Among his [Caracalla's] works at Rome he left the magnificent Baths which bear his name, the *cella solearis* of which architects say cannot be imitated in terms of construction. For it is said that lattices either of bronze or of copper were placed over (or under) to which the whole vault is entrusted and the span is so great that experienced engineers say it could not have been done.

The definitive work on this passage was done, once again, by Professor Janet Delaine.[13] As she notes in her paper, several words in it appear nowhere else, including "solearis." It appears that Brödner may have misinterpreted this as "solar" or "sunny" (the word, according to Delaine, seems to have a completely different root). Together with an inaccurate translation of the rest of the passage, this might suggest lightweight construction used in place of a roof. Delaine interprets it as the lightweight framework used to support a roof. In any event, there seems very little here to give confidence that the Baths had mirrors on the top.

So we had a mistranslated passage that was cited, with misgivings, by Plommer, who was cited, again with warnings, by MacKendrick (who issued no warnings), who was cited by Roth, who was cited by Wikipedia (which gives the wrong room), which was cited by the Internet.

So, after all is said and done, we have no actual reference to mirrors being used to direct sunlight into a Roman building.

We can infer that mirrors were used to direct sunlight into buildings, at least temporarily, from the existence of "magic mirrors" in China and Japan. These are metal mirrors with no obvious surface figure that, when used to mirror sunlight onto a dark wall, reveals an image. The mirrors can certainly be traced back to the 9th century CE. They possibly date back to the 2nd century BCE. This novelty device indicates that people certainly did redirect sunlight into buildings, for at least such a use, but doesn't support belief in redirected solar illumination as a regular practice.

The first definite cite I have found about someone making a permanent construction for bringing sunlight into a building is the work of the mathematician and natural philosopher Willems Gravesande (1688–1742), who invented what he called a *porte lumiere* (French for "light door"). In its full form, it consisted of a mirror and a condenser to redirect sunlight into a darkened room to provide light for experiments. It was reinvented by Thomas Drummond (1797–1840), who called it a *heliostat*. Both terms came to be used, with *porte lumiere* being a device with no automatic drive to adjust the mirror, while a *heliostat* came to be one in which a clockwork mechanism kept the first mirror directing the sunlight in the proper direction to keep the output light direction steady. Both devices were used in the second half of the 19th century for instructional purposes, before the development of reliable and bright artificial electric lighting for experiments.[14]

Even with a clockwork drive, however, it was touchy and inefficient to try to send sunlight into structure using mirrors. As the cities became larger, with taller buildings blocking out sunlight, people did look for devices to help direct light into buildings. But at first these were passive devices, using glass prisms instead of mirrors. Glass brick started to be used more often, and prismatic bricks that could take sunlight and refract it into rooms started to be used. Thick glass blocks set into iron frames were placed on sidewalks to send sunlight into basements, the bottoms of the glass blocks being formed into prism shapes so that the light did not go straight down.

It's only now, in the 21st century, with cheap and small programmable machines at our disposal that heliostat systems are starting to be used as practical architectural elements to send sunlight into buildings. Once the sunlight has been directed along a critical axis by the programmed heliostat, further internal mirrors of high reflectivity can carry the light further into the interior of the building with high efficiency. What had been a fancy frillip has, with computing power and improved optics, become a practical reality.

The ultimate embodiment of this has been the use of huge mirrors to redirect sunlight into towns in deep valleys that might not see direct sunlight for weeks. Thus far this has been done for two towns—Rjukan in Norway and Viganella in the Italian Alps. Viganella constructed its mirror system in 2006, using a single mirror measuring 8 m × 5 m to direct light into the town square. Rjukan built its mirrors in 2013. Rjukan's three mirrors, each measuring about 17 m², cast light on the Rjukan town square for a period of about two hours in January, when the village is least lit. These lights are intended to help relieve seasonal depressive disorders due to sunlight deprivation.

Even more ambitious are ideas for orbiting "sun mirrors" to direct light into areas needing it. NASA wrote a report on the idea in 1982. The Russian space program not only studied the idea, but briefly tested it with the satellite *Znamya* ("Banner") in 1993, using a 65 foot diameter aluminized mylar reflector to send a 2.5 mile wide beam down to the earth. The Chinese are reportedly considering an "artificial moon" for night-time illumination to be launched in 2020.

Notes

1. Norman Lockyer, *The Dawn of Astronomy: A Study of the Temple-worship and Mythology of the Ancient Egyptians* (New York: Macmillan and Company, 1893).
2. Charles-François Dupuis, *Origine de tous les cultes ou religion universelle* (Agasse, 1794).
3. Stephen R. Wilk, Chapter 2, "The Solar Weapon of Archimedes," in *How the Ray Gun got Its Zap!* (New York: Oxford University Press, 2013).
4. https://tvtropes.org/pmwiki/pmwiki.php/Main/LightAndMirrorsPuzzle (accessed November 9, 2019).
5. *Mythbusters* Episode 169, "Let There be Light," broadcast June 22, 2011. https://mythresults. com/let-there-be-light (accessed November 9, 2019).
6. See Chapter 1 in this book.
7. https://en.wikipedia.org/wiki/Baths_of_Caracalla (accessed November 9, 2019).
8. Leland M. Roth, *Understanding Architecture: Its Elements, History, and Meaning* (New York: Icon Editions, 1993).
9. Paul Lachlan MacKendrick, *The Mute Stones Speak: The Story of Archaeology in Italy* (New York: WW Norton & Company, 1983).
10. Hugh Plommer, *Ancient and Classical Architecture* (London: Longmans, Green, and Co., 1956), 341–344.
11. Erika Brödner, *Untersuchungen an den Caracallathermen* (Berlin: W. de Gruyter, 1951).
12. Translation from this site: http://penelope.uchicago.edu/Thayer/E/Roman/Texts/Historia_ Augusta/Caracalla*.html
13. J. Delaine, "The 'Cella Solearis' of the Baths of Caracalla: A Reappraisal," *Papers of the British School at Rome* 55 (1987): 147–156.
14. On the history and use of these devices, see Amos Emerson Dolbeare, "Projections for the Schoolroom," *New England Journal of Education* 3, no. 8 (January 15, 1876): 28; and *The Art of Projecting: A Manual of Experimentation in Physics, Chemistry, and Natural History with the* Porte Lumiere *and Magic Lantern* (Boston: Lee and Shepard, 1892). The latter is online at https://books.google.com/books?hl=en&lr=&id=k2sGAQAAIAAJ&oi=fnd&pg=PA12 &dq=portelumiere&ots=PzlHiidUog&sig=cdAWuclA_WswhOfauDHdA0aUPqE#v=one page&q=portelumiere&f=false. Dolbeare was a professor at Tufts University.

18
Deck Prisms and Vault Lights

You can see them in museum shops and nautical stores, in gift boutiques and catalogs. They've been made in miniature as earrings and as desk curios. Or, in full size, they sit atop illuminated bases or swing in specially made chain mounts. They're deck prisms, large pieces of cast glass in the shape of hexagonal pyramids. The purest ones are in slightly bluish "clear" glass, but there are versions in blue cobalt glass and in other colors—impractical for their original use, but more pleasing to the eye, especially when illuminated (Figure 18.1).[1,2]

As the documentation accompanying these make clear, these are all inspired by the "deck lights" or "deck prisms" found aboard the *Charles W. Morgan*, a whaling ship built in 1841 and maintained at Mystic Seaport in Connecticut. They started selling replicas in the gift shop there, and they've been spreading ever since.

The idea is simple and elegant—these prisms were inserted into hexagonal holes chiseled into the deck, and they provided light to the deck below. The somewhat rippled sides of the hexagonal prism served to refract and diffuse the light, instead of letting it fall into a mere circular patch, as a disc of thick glass set into the deck might do. It was watertight, but let in light, and didn't require the use of lamps or candles, and so avoided the risk of fire onboard. Most of the newly purchased ones go on display by themselves, but some people are installing them, or other models on actual ships, where they add a touch of nostalgia (and, one suspects, a chance of painful bumps on the head).

Where and when did this idea originate? A survey of websites reveals the curious consistency of the Internet—the result of people cribbing from the same websites or from each other, without checking for independent corroboration—*the concept dates back to the 1840s*, the sites say, *and possibly centuries back before that*. But the claims are suspicious and vague. None of them cites a source, and they smack of assumption and wishful thinking. That one certain decade—the 1840s—looks as if it is suggested by the date of the launching of the *Charles W. Morgan*. But there is no evidence that the Morgan was launched with deck prisms in place.[3,4,5]

Searching through books, I find a couple of references to deck prisms in historical romances. Kerry Lynne's 2013 novel *The Pirate Captain: Chronicles of a Legend* makes frequent mention of the "deck prism" and the greenish-blue light it casts. The story is set in the 18th century, so clearly the author (trained as a historian, and a sailor herself) either thought such deck prisms were used then or stretches the truth for some picturesque scenes. Similarly, Delle Jacobs, in her 2009 novel *Sins of the Heart*, set in 1813, writes about the "dim light coming in through the deck prism." It makes for a good scene, but is it at all believable? Recent romance novels aren't a good source for historical fact.

So we return to the original question: where and when did this idea originate?

Figure 18.1 Reproduction of the deck prism used on the whaler Charles W. Morgan. The prism as shown is "upside down." In use, the prism was inserted into a hexagonal opened cut into the deck, with the flat "base" flush with the deck and the pointed end protruding into the ceiling of the deck below.

Image courtesy of Mystic Seaport, Mystic, CT.

At this point, I am indebted to Paul O'Pecko, historian at Mystic Seaport, for a critical nugget of information. The original name for a thick glass piece set into the deck or the wall of a ship, it turns out, is not *deck light* or *deck prism*, but *bullseye*. And with that revelation comes a vast outpouring of information, and a possible early history of the idea. I hope the reader will pardon some etymology and some speculation on my part, and a side excursion into the world of glass technology. [6]

For most of us, the first association of "bullseye" is the target, used in archery, pistol shooting, darts, and other recreations. But this use seems to be relatively late, dating from 1833, according to the *Oxford English Dictionary*. A search using the Google resources turns up earlier usages, including one from 1813 in this sense.

But an even earlier use of bullseye is to mean a glass lens, or a thick glass window. "Bullseye" lantern, meaning a glass lantern with a thick concentrating window, is cited by at least 1800. One finds reference to "bullseye" spectacles and "bullseye" meaning lenses from the early 19th century.[7,8] Why should "bullseye" be used in this context? To understand why, we must look at the early means of manufacturing window glass.

One popular method was to gather a gob of molten glass on the end of the pontil, the rod used to handle the glass and for glass blowing, and to rapidly spin this to form the glass into a disc. (Another method was to use the pontil as a blow-pipe and blow air into the gob, turning it into a sort of thick glass balloon, which was broken at the end and spun, as before, into a flat disc). The disc was kept in motion until the glass cooled sufficiently to retain the shape. When sufficiently cool, the large disc of glass, which had nearly parallel front and back faces, was cut into pieces to be used for windows, especially the diamond-shaped panes that were bound together with leaden strips to form the English diamond-paned windows.

The only portion that could not be used for this purpose was the very center, about which the glass disc was spun, and to which the pontil was attached. This section was necessarily thicker than the rest, a somewhat conical piece of glass without parallel front and back surfaces, and with an opaque center where the pontil was attached. It was unsuitable for window glass. Because of its shape, however, it was perfect to use as raw material to make lenses from. One name for this portion was the *crown*, and from it one could make crown glass lenses. We still call low-index soda-lime glass a "crown" glass (in distinction to the higher index "flint" glasses that were made of lead-based starting materials). "Crown" glass was also used to refer to windows made from this section of the glass. Not useful for windows that must be optically clear for viewing, it was nevertheless useful for places where viewing was not important, but the admission of light was. Hence, crown glass was often used in places like transoms. "Crown" glass was also called "bullseye" glass, though. Such circular pieces were called "bullseyes," as were the lenses sometimes made from them. Lanterns using such lenses as light concentrators were "bullseye" lanterns.

Why? It's significant that another name for these lenses and pieces of glass was *bullion*. "Bullion" is a word with an interesting and confused history, but "bullion" glass seems to be related to "bullion" referring to bars of precious metal, like gold or silver. Both can ultimately be traced back to a French word meaning "molten" and related to "boil" (although there are competing etymologies). It surely cannot be coincidence that the similar-sounding "bullion" and "bullseye" mean the same thing. There is a characteristic tendency in British English to take foreign words and convert them into something more familiar and English. Thus, "chartreuse" became "charterhouse." It would not be at all surprising if the English glassmakers (who were noted for this type of glass) transformed the French term "bullion" into the English term "bullseye."

The use of the term "bullseye" then extended to items made from it, including lenses. The concentric circular patterns in the glass then provided the impetus to name the target made of concentric circles also a "bullseye"—which would thus have no direct bovine roots at all, although later users of the term might have assumed that and continued to use the term under the assumption that there was such a relationship.

But we are getting afield from our main point, which is the use of the term "bullseye" for thick pieces of glass set into the deck of a ship or into the sides to admit light without letting in water.

At this point we enter into likely surmise. Just as the despised, unwanted, and therefore inexpensive bullseyes were used for transom lights, some inventive and enterprising ship's officer must have seen that the thick pieces of glass could be put to use for the same purpose by cutting holes in the deck and inserting the thick pieces of glass to provide below-deck illumination, or putting them in ship sides. Set firmly in place with oakum and sealed with putty, they would not leak. Being thick, they would not break under the pressure of seamen's feet. On warships, such thick pieces of glass could provide an extra barrier between illuminating flames and the gunpowder in the powder magazine.

No documentation supports this series of conjectures, but they do provide a plausible explanation for the use of thick glass as deck lights, and explain why they have the same name as the central portions of crown glass. The earliest use we have of "bullseye" for such a ship's light is 1829, with many uses in the early 1830s. A use of "bullseye" for the light in a powder magazine comes from 1833.

A very similar idea seems to have been independently developed on land, for quite a different purpose. In the growing cities, paved sidewalks were coming to be used, but they often covered basement rooms that also needed illumination. Thick pieces of glass, specially cast and carefully shaped (unlike the irregular crown glass bullseyes) were set into cast iron frames that were set in the sidewalk. These admitted light to the spaces beneath, but were thick enough to be walked on without damage. Such structures were called vault lights.

Some sites credited Thomas Hyatt for inventing the concept in 1845, but Hyatt himself did not claim to have invented the idea. [9]The earliest case I can find is a March 8, 1834, patent by Edward Rockwell of New York City. The patent is so old that the patent document online is handwritten. The patent drawing shows an annular metal frame (probably cast iron) set with Victorian filigree, surrounding a circular glass disc that is "stepped" in bullseye fashion. A cross section shows that the disc is flat on the upper surface but is a sort of stepped cone on the lower surface, allowing many surfaces for the light to scatter into the space below. The ring reads *Rockwell's Patent Vault Light—New York*. Magazine records show that he was awarded a silver medal for the invention. [10]

Thomas Hyatt's "Illuminating Vault Cover" was patented on November 12, 1845, #4266, and consists of many circular thick pieces of glass set in a cast iron framework. The advantage of his construction, he notes in his patent, is that the many smaller pieces of glass are less liable to fracture than the single plano-convex piece of the unnamed Rockwell. Hyatt's illustration shows that his glass pieces were simply thick cylinders, so there was no refracting or scattering structure to help conduct light into the space illuminated. Nevertheless, Hyatt, too, won a silver medal in 1855 from the American Institute in New York. [11]

Over the next 70 years, many others proposed designs for such vault lights. One of the more interesting for our inquiry is patent #17096, granted on April 21, 1857, to George R. Jackson of Rye, New York. Jackson's vault lights are of several possible shapes, [12]set into a cast iron frame, conspicuous among which are hexagonal pyramids, just like the Mystic Seaport deck prisms (Figure 18.2).

Jackson states in his patent that

My improvement in illuminating vault-covers consists in the peculiar shape of the glasses employed by me for closing the apertures in the metallic portions of said covers—viz, glasses in the form of an inverted pyramid, or other equivalent polygonal form which will, by reflection and refraction, laterally diffuse the descending rays of light uniformly throughout an apartment, and especially the upper portion thereof —substantially as represented in Figure 1, of the accompanying drawings.

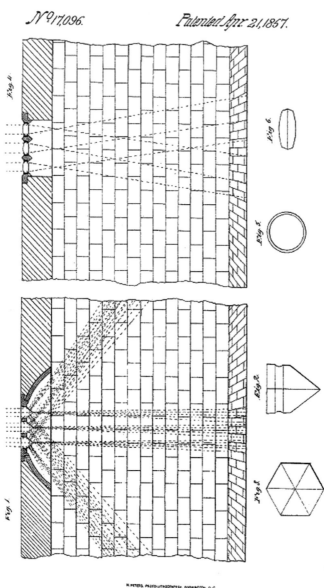

Figure 18.2 Illustration from Jackson's 1857 patent for vault lights, showing the hexagonal form of the prism used, identical to the deck lights on the Charles W. Morgan. US Patent Office

In Jackson's patent we have, so far as I have been able to find, the first expression of the "deck prism" shape, and for exactly the usually stated reasons. Jackson recognizes the earlier work by Hyatt and Rockwell, calling them by name.

Jackson, too, won a silver medal at the 28th annual Fair of the American Institute in October 1856. It seems to have been some sort of rite of passage for vault light inventors. Who was Jackson?

He was born in New York City on June 4, 1811, and as a boy was apprenticed to a "whitesmith"—an iron worker who does fine finishing work. He founded his own firm on Centre Street in New York City in 1839 and then entered into a partnership with a Mr. Cornell. Cornell and Jackson lasted until 1846 or 1848, when Cornell died. He partnered for a time with L. Taylor, who also died shortly afterward. He then joined James J. Burnet to form the Excelsior Iron Works at 340-352 East 14th Street. The building was destroyed in a fire in 1869. [13]

Two of his sons joined the firm, along with nephews, and some plaques, like the one still in place in the historic Smith, Gray, and Co. building in New York, observe that the iron frontwork was done by "George R. Jackson and Sons Foundry." Perhaps the foundry was considered distinct from the company, which was named "George R. Jackson, Burnet, and Co." by this time. Jackson was a member of the East River Association, an industrialist group promoting the interests of manufacturers. He died in September 1870 after a long illness. He lived at 85 East 10th Street. His partner James J. Burnet survived him by 20 years.

An advertisement in the *New York Times* for the Excelsior Works in 1859 says that it was founded in 1839, which is perhaps literary license for Jackson's being in the cast iron business since that year, even if under a different name.

The vault light was exactly the sort of thing that Jackson would be interested in, since it provided a market for his cast iron. Jackson filed patents for many iron devices. His patent for the vault light came when he was 47. James J. Burnet is listed as a witness on the patent. Jackson's patent #18851 also used pyramidal prisms. [14] An ad in the *New York Times* in 1859 says that the Excelsior Iron Works manufactures "patent vault lights."

Intriguingly, Thaddeus W. Hyatt filed another patent, #21050, granted on July 27, 1858, for a vault light using "an inverted pyramidal, polygonal, or conical form ... for the purpose of producing a wide-spread and perfect diffusion of the rays of light which might pass through said cover into the apartment beneath." The description is so similar to Jackson's, right down to the wording, that you have to wonder why a separate and later patent was granted. Even more intriguing, the patent, although said to be invented by Hyatt, was assigned to George R., Jackson. Perhaps we see here the outcome of a patent battle.

Some websites, speculating on the origins of deck prisms and vault lights, suggest that the sidewalk devices were inspired by ship's deck lights. But it appears to have been the other way around. The earliest reference Mystic Seaport has located is an advertisement from 1855 referring to deck prisms. But there is a reference to them four years earlier. The *Official Description and Illustrated Catalog of the Great Exhibition of the Works of Industry of All Nations* lists the invention of vault lights for ships in 1851 (Vol. II, p. 6489) by James Barlow of 14 King William Street, Mansion House, City of London.

Barlow was a furnishing ironmonger, born in 1797, who was also a prolific inventor. Another of his inventions listed in the Official Description for 1851 is a siphon-tap for kegs. An illustration of this appears in George Meason's *Official Illustrated Guide to the London and South-West Railway* from 1858. Barlow died in 1862 in Leatherhead.

I haven't been able to find an illustration of Barlow's vault light, but it probably wasn't in the form of an inverted pyramid. It seems unlikely that the ship's prism was invented first, and certainly not in that pyramidal form. If it were, the earliest patents for vault lights would have copied that design, and at least one of them would have mentioned the inspiration.

Some sites I have encountered claim that another name for these devices is "dead lights." This seems counterintuitive, and I suspect there is an error here. "Dead lights," as the name implies, were panels placed over cabin windows to protect them from breakage during heavy weather, and so "killed" the light coming in. At some point someone hit upon the idea of putting a heavy glass sphere into the panel, so that not all light would be lost, and the thick glass was resistant to breakage. But these panels were vertical ones placed over vertical windows.

I have also come across historical romance novels that speak of pyramidal deck prisms being used in the 18th century or even earlier, but one ought not to go to novels for historical information. This appears to be a case of assumption of great age for these devices on the part of the writer. Or literary license.

Notes

1. See, for instance, this site: "The earliest deck prisms with provenance are from the 1840s. Presumably they were used earlier, but how much earlier is unknown; the origin of the idea is lost, and glass is difficult to date." http://www.glassian.org/Prism/Deck/ (accessed January 25, 2015). This site, *Marine Insight* for June 19, 2012, makes identical claims: http://www.marineinsight.com/marine/marine-news/headline/what-is-a-deck-prism-on-ships/ This site, *Moby Cargo*, which sells such lights, makes similar claims: http://www.mobyscargo.com/original-small-green-deck-prism. *WiseGeek* repeats the claim: http://www.wisegeek.com/what-is-a-deck-prism.htm#didyouknowout

2. 1818 use: "Journey to Lake Mánasaróvara in Un-dés," *The Asiatic Journal and Monthly Register for British India and its Dependencies* (March 1818): 236 and 237. https://books.google.com/books?id=cpFNAAAAcAAJ&pg=PA237&dq=%22bull%27s+eye%22&hl=en&sa=X&ei=vYatVKOlDMukermtgrgF&ved=0CCoQ6AEwADgU#v=onepage&q=%22bull's%20eye%22&f=false (accessed December 25, 2015). Another 1818 citation is "Monthly Register – British Chronicle – Archery" in *The Edinburgh Magazine and Literary Miscellany* 3 (August 1818): 182. https://books.google.com/books?id=q9U5AQAAMAAJ&pg=PA182&dq=%22bull%27s+eye%22&hl=en&sa=X&ei=S9quVL-mKMbksASslIHoAw&ved=0CEEQ6AEwBTgo#v=onepage&q=%22bull's%20eye%22&f=false (accessed December 2014). The earliest usage I found, from 1813, is in "Observations Relative to the Near and Distant Sight of Different Persons" by James Ware Esq. in the *Philosophical Transactions of the Royal Society of London* 103, Part 1, 31–50, with the quote on p. 32. The paper was read on November 19, 1812, which actually makes the citation a year earlier. https://books.google.com/books?id=qpRJAAAAYAAJ&pg=PA32&dq=%22bull%27s+eye%22&hl=en

&sa=X&ei=ztyuVJnrM4i1sQScs4BA&ved=0CFMQ6AEwCDhG#v=onepage&q=%22bull's%20eye%22&f=false (accessed December 2014).

3. A use from 1800 appears ion Charles Reade's novel *Hard Cash: A Matter-of-fact Romance* on p. 254. https://books.google.com/books?id=9Lw0AQAAMAAJ&q=%22bull%27s+eye%22&dq=%22bull%27s+eye%22&hl=en&sa=X&ei=U9yuVPi8JuGHsQSToDIDw&ved=0CFkQ6AEwCTgy (accessed December 2014).

4. One reference on this method: *Five Black Arts: A Popular Account of the History, Processes of Manufacture*, edited by William Turner Coggeshall. https://books.google.com/books?id=TU4yAQAAMAAJ&pg=PA211&dq=%22crown+glass%22&hl=en&sa=X&ei=kkOsVIeZCITbsASoh4A4&ved=0CD8Q6AEwBjgK%20-%20v=onepage&q=%22crown%20glass%22&f=false#v=snippet&q=%22crown%20glass%22&f=false

5. http://en.wikipedia.org/wiki/Charterhouse

6. It is significant that neither Samuel Johnson's 1755 *Dictionary of the English Language* nor Noah Webster's 1806 *A Compendius Dictionary of the English Language* contains the word *bullseye*, although Johnson has *bull eyed*.

7. William Falconer, *New Universal Dictionary of the Marine*, 2nd ed. The first edition of 1769 significantly does not include the term. I am indebted to Paul J. O'Pecko, director of Collections and director of the G. W. Blunt White Library at Mystic Seaport, for this and the next reference.

8. Henry Barnet Gascoigne, *The Path to Naval Fame with an Index of Nautical Terms and Phrases*, 2nd ed.

9. One place that gives Hyatt the credit is Deitrich Neumann in "Prismatic Glass," Chapter 23 in *Twentieth Century Building Materials: History and Conservation*, ed. Thomas C. Jester (1995), 157. Another is a publication of the US Park Service: Cas Stackelberg and Chad Randl, *Preservation Tech Notes – Historic Glass Number 2: Repair and Rehabilitation of Historic Sidewalk Vault Lights*. November 2003. http://www.nps.gov/tPS/how-to-preserve/tech-notes/Tech-Notes-Glass02.pdf. Another is the webpage *Hidden City Philadelphia – Diffused Down Below; Philadelphia's Lost Vault Lights*, June 19, 2013, by Sam Robinson. http://hiddencityphila.org/2013/06/diffused-down-below-philadelphias-lost-vault-lights/ (accessed December 2014).

10. *Niles Weekly Register* 47 (1834), p. 128. https://books.google.com/books?id=Y7ARAAAAYAAJ&q=%22vault+light%22&dq=%22vault+light%22&hl=en&sa=X&ei=JJOIVNC-OraAsQSOm4KwBA&ved=0CEwQ6AEwBzg8

11. https://books.google.com/books?id=veU8AQAAIAAJ&pg=PA93&dq=%22vault+light%22&hl=en&sa=X&ei=uZCIVMi1IciIsQTvjIHoAw&ved=0CFQQ6AEwBzge%20-%20v=onepage&q=%22vault%20light%22&f=false#v=snippet&q=%22vault%20light%22&f=false. See p. 93 of *Transactions of the American Institute of the City of NY Annual Report* 1855 (publ. 1856).

12. "George R. Jackson Awarded Silver Medal … for His Vault Light at the 28th annual Fair of the American Institute, October 1856, Manufacturing and Mechanical Department: Building Materials. To G.R. Jackson & Co. of 201 Centre Street, for best Vault Light. A Silver Medal (The Hydeville Marble Works took the gold, for their Best Slate Stone Mantels, Marbleized)," p. 130 in *Transactions of the American Institute of the City of New York* (1857): https://books.google.com/books?id=tBFAAAAAYAAJ&pg=PA130&dq=%22vault+light%22&hl=en&sa=X&ei=uZCIVMi1IciIsQTvjIHoAw&ved=0CD4Q6AEwAzge%20-%20v=onepage&q=%22vault%20light%22&f=false#v=snippet&q=%22vault%20light%22&f=false

13. For references on Jackson: Barbaralee Diamonstein-Spielvogel, *The Landmarks of New York, Fifth Edition: An Illustrated Record of the City*, 230. He was a member of the East River Association, an industrialist group. See Iver Bernstein, *The New York City Draft Riots: Their*

Significance for American Society and ... https://books.google.com/books?id=SF0rWO4y-JYC&pg=PT244&dq=%22George+R.+Jackson%22&hl=en&sa=X&ei=X7yHVM3aKYuUyAS3jIDwCA&ved=0CDYQ6AEwBTgK#v=onepage&q=%22George%20R.%20Jackson%22&f=false. Landmarks Preservation Committee (NYC) Report: http://www.nyc.gov/html/lpc/downloads/pdf/reports/smith_gray_bldg.pdf More information on him here: http://www.waltergrutchfield.net/burnet2.htm

14. "Hyatt Assigns Vault Light patent to George R. Jackson": https://books.google.com/books?id=4GhHAQAAIAAJ&pg=PA177&dq=%22vault+light%22&hl=en&sa=X&ei=9ZyIVP3XOa-1sQTalILIAw&ved=0CEsQ6AEwBA#v=onepage&q=%22vault%20light%22&f=false; *Congressional Series of United States Public Documents Executive Documents Printed by Order of the House of Representatives during the Second Session of the 35th Congress, 1858–59*, Vol. 986; Vol. 1010, p. 176, patent #21,050.

19

Forty-Four Fewer Shades of Gray

Some time ago I was looking at a color checker chart—that rectangular chart having 24 squares arranged in four rows of six (originally manufactured by the Macbeth company, which has, through mergers, been acquired by X-Rite). This is very widely used to check on the response of cameras, films, and, increasingly, camera systems to a standardized series of colors. The third row down contains the three additive and the three subtractive primary colors, which are undoubtedly responsible for the choice of six horizontal elements. The two upper rows contain a selection of colors intended to be representative of things commonly photographed—dark and light skin, blue sky, the green of many plants, and so on. The bottom line contains six colorless squares that represent a range of grays, from completely white to completely black. And this was the line that attracted my interest.

Were I putting together such a chart, my first impulse would be to have the grays advance in regular steps of reflectivity—0%, 20%, 40%, 60%, 80%, and 100% reflecting. This shows how naïve I am. The actual reflectivities, as set down in the 1967 paper proposing the chart,[1] are as follows:

3.1%,
9.0%,
19.8%,
36.2%,
59.1%, and
90.0%.

The steps are neither uniform in size nor obvious.

Well, if the steps aren't equally spaced in reflectivity, then surely they are uniform steps in optical density, right? Again, no. The corresponding densities are

0.046,
0.228,
0.441,
0.703,
1.046, and
1.509.

The first three squares are roughly uniform in step size, about 0.22 apart. But the next one, at 0.703, is 0.26, and the differences get larger from there.

The truth is, as photographic science and color science people reading this are already aware, that the steps are chosen on a different basis. As C.S. McCamy et al. explicitly state in their paper defining the shades state, "The series is equally spaced on the Munsell system." And, indeed, using the Munsell system of color identification, all of the squares in the bottom row have zero "Chroma" but differ by steps of 1.5 in "Value," being

9.5,

8,

6.5,

5,

3.5, and

2.

But this simply drives the question one step further back—McCamy et al. may simply have taken equally stepped gray squares from the Munsell color pack, but where did Munsell get those values from? Each step in the Munsell "Value" designation is supposed to represent an equally sized step of gray as perceived by humans, the result of the interaction among the (standardized) light source, the pigmented color squares, the human eye, and the color-processing machinery of the human brain. Since the human visual system is part of the process, this is the ultimate in subjective evaluations, and subjective measurements are notoriously variable. Certainly we know from experience that sensory results vary from person to person, and we ought to expect one person's idea of equal-sized steps to gray to differ from some other randomly chosen person's notion. But people are, statistically, pretty much the same, and in statistics lies our salvation. Taking the results from a larger number of subjects, we expect these human differences to average out.

Again, were I to set up a procedure for determining the reflectivities of six equally spaced shades of gray, as determined by the average human being, my thought would be to generate a large number of gray test squares of known reflectivity, and to get a large group of volunteers (say, about 100), chosen to represent a broad spectrum of humanity, and have them sit at a color-viewing cabinet with undistracting gray walls illuminated by a standard light source and have them select what they thought were six equally spaced shades of gray. To guarantee that they used the same end points, I might have them use predetermined "black" and "white" endpoints. Then you simply tally the results, taking the averages of each gray level, which ought to fall in some sort of normal distribution, ideally at the peak.

It probably won't surprise you to learn that this is not the way they did it at all. There was a long history to the determination of human sensations to various levels of gray that shaped the experimental process.

Historically, the problem has been addressed from the other side—What difference in reflectivity produces a uniform change in the sensation of grayness? People had been trying to solve that problem since at least the 18th century. Pierre Bouguet was a professor of hydrography, born at Le Croisic in Brittany. He was also a mathematician, geophysicist, and astronomer, among other things. It was in this last office that he discovered the Beer-Lambert law, showing in his 1729 book *Essai d'Optique sur la Gradation de la Lumiere* that the intensity dropped exponentially with distance traversed through a medium. It was undoubtedly his interest in the variation in light intensity and the ability of the human eye to detect differences in it that led him to investigate what the just noticeable difference detectable by the eye was. His result, published in 1780, two years after his death, in the similarly named book *Traité d'optique sur la gradation de la lumière*.[2] He stated that the just noticeable difference was proportional to the brightness of the source, and that in the case of his own eye the difference in brightness was 1.5%.

This general rule was confirmed by the work of many scientists studying human vision, including Francois Arago, Carl August von Steinheil, Antoine-Philibert Masson, and others (who obtained different constants of proportionality). The name for the general law for it derives from the work of the German physician Ernst Heinrich Weber and his student Gustav Theodor Fechner. The Weber-Fechner law refers to two somewhat different formulations, one in differential form and the other essentially its integral, in the form

$$V = c \log R + k$$

where V is the sensation, R is a measure of the stimulus (and is usually the Reflectance in visual studies), and c and k are constants. (The Weber-Fechner law of logarithmic response to stimulus has been applied to other senses since—touch, sensation of weight, etc. But it was first applied by Weber and Fechner to visual intensities.)

This simple law, however, soon found complications it appeared that the fraction corresponding to the just noticeable difference, often called the "Fechner fraction," wasn't really a constant, but varied slowly with intensity.

The Belgian physicist Joseph Antoine Ferdinand Plateau (who, among other things, invented the phenakistiscope, a stroboscopic forerunner of the cinema) suggested a different relationship, now known as Plateau's law:

$$V = kR^c$$

where V and R have the same definitions as earlier, and k and c are again constants (although completely different ones). There were thus, by the end of the 19th century, two different equations relating sensation V to stimulus R, one logarithmic, and the other exponential. Clearly both could not be correct. It turns out, in fact, that neither properly accounts for the relationship between the reflectance of light and the human response across all circumstances.

When painter and color theorist Albert Henry Munsell started developing his notation for colors around 1900, he looked for a reliable system and could find no consensus. He provisionally adopted Plateau's law, invented his own photometer to use for measurements, and continued to make measurements clarifying this (and other aspects of his color scheme) until he died in 1918. He had established the Munsell Color Company in 1917 to perform the work of working out and marketing his ideas on standardized colors, and after his death the company was run by his son, Alexander Ector Orr Munsell. The younger Munsell, with scientists from the Munsell Company, continued to research the specification of color and to publish their findings.

They were able to produce an experimental curve showing the average relationship between the reflectance of the target and the "Value" of the gray as perceived by the subject.[3] In some cases it could be stated as an integral number of just noticeable differences. The exact shape of the plot has the general form of a fractional power law, as the Plateau model might suggest, but it does not quite fit a simple square root or cube root formula, and over the years a variety of formulas have been implemented, generally in the form of a fractional power law with one or more correction terms.[4] Using any of the formulas or interpolating between averaged experimental points allows one to give a series of grays separated by steps of equal sensation. Munsell et al. give such a series of reflectances for a scale of 10 steps. They undoubtedly chose 10 steps because of our use of a decimal system, but you can construct a system for any number of steps.

So at last we can answer the question of how the steps of equal stimulus were constructed—the Munsell Company did indeed use a battery of observers in a controlled viewing booth. The difference between using six steps and ten steps isn't a problem once you have a curve relating reflectance to response. The only thing that can raise concern is the number of test subjects. In the Munsell, Sloan, and Godlove 1933 paper, they used different numbers of subjects,[5] with a maximum of 16 for one test. That's far below my suggested 100 subjects. I don't think any of the series of tests by any of the experimenters rise above that set of 16 test subjects.

Of course, repeated testing by multiple workers on varying numbers of subjects confirmed the general shape of the curve, even if no one test featured more than 16 subjects, so the overall number of measurements confirming the curve is much higher. Most papers since the 1940s take the form for granted and produce their modifications of the formulas from earlier data.[6,7,8]

Notes

1. C. S. McCamy, H. Marcus, and J. G. Davidson, "A Color Rendition Chart," *Journal of Applied Photographic Engineering* 2, no. 3 (Summer 1976): 95–99.
2. Pierre Bouguer, *Traité d'optique sur la gradation de la lumière* (Guerin & Delatour, 1760).
3. A. E. O. Munsell, L. L. Sloan, and I. H. Godlove, "Neutral Value Scales I. Munsell Neutral Value Scale," *Journal of the Optical Society of America* 23, no. 11 (November 1933): 394–411.
4. You can find 11 of these given in the Wikipedia page of "Lightness"—https://en.wikipedia.org/wiki/Lightness (accessed November 12, 2018), including the current CIELAB formula that is almost a cube root with offset. There are many others given in I. H. Godlove's 1933 paper.
5. Munsell, Sloan, and Godlove, "Neutral Value Scales I. Munsell Neutral Value Scale."
6. I. H. Godlove, "Neutral Value Scales II. A Comparison of Results and Equations Describing Value Scales," *Journal of the Optical Society of America* 23, no. 12 (December 1933): 419–425.
7. Sidney M. Newhall, Dorothy Nickerson, and Deane B. Judd, "Final Report of the OSA Subcommittee on the Spacing of the Munsell Colors," *Journal of the Optical Society of America* 33, no. 7 (July 1943): 386–418.
8. Elliott Q. Adams, "A Comparison of the Fechner and Munsell Scales of Luminous Sensation Value," *Journal of the Optical Society of America* 6, no. 9 (November 1932): 932–939.

20
Barrel and Pincushion

Among the aberrations, the least discussed is what is usually the last one, distortion. The third-order term for distortion is independent of the pupil coordinates through which the rays pass and depends only upon the object height. If the coefficient for the distortion is negative, the image points are closer to the center than they should be, by an amount that increases the further the point is from the center. The result is barrel distortion. The name derives from the form of a rectangular grid imaged by a lens having such aberration, which has the bowed-outward sides that make it resemble a barrel. (The top and bottom bow outward, too, like a barrel filled with something generating too much pressure, but the naming ignores that.)

If, on the other hand, the sign is positive, the distortion makes the points seem to be too far away from the center, again varying more with increased distance from the center. When this happens, we have pincushion distortion, named because the sides of an imaged square are concave outward, with the corners being extended in sharp points. This, it is claimed, makes it resemble a pincushion.

Except that, to me, it doesn't. I grew up seeing pincushions that looked like cloth tomatoes, sometimes with inexplicable strawberries attached. I never, in real life, saw a pincushion that looked like one of those pointy-cornered squares. Clearly, the people who named the optical effect were familiar with something that had passed me by. Thus another way we learn about the past—from the social conventions fossilized in our language.

Sewing needles date back to the Neolithic, and I assume that pins do as well. Surely people were doing something with those easily lost (and painful to find) clothing-making items. The Internet sites I've visited seem to indicate that they were stored in etuis (that word beloved by crossword puzzlers) and similar cases, and that pincushions first appeared in the 18th century in France. I suspect that long before that, people were finding convenient things to stuff needles and pins into. In any event, by the 1700s people were sewing together two squares of cloth and stuffing them heavily with some sort of padding so that the needles stuck into the resulting construction wouldn't poke out the other side. This stuffing made the "pillow" thicker in the middle, pulling in the sides, but not affecting the corners as much. Viewed from above, the pincushion looked like a rectangular grid viewed through a lens with positive distortion. Voila!

Where did the tomato-and-strawberry pincushions of my youth come from? Various sewing websites give the same explanation, saying that these are the result of a Victorian custom of placing a tomato on the mantelpiece to ensure prosperity and to ward off evil influences. Eventually the prone-to-decaying vegetable was replaced by a longer-lasting cloth imitation. Presumably, because it was handy and suited to the purpose, people started using the ubiquitous good-luck piece as a handy holder for pins and needles. Eventually they started deliberately sewing tomato-shaped pincushions, attaching a strawberry-shaped mini-cushion filled with emery or some other abrasive to sharpen pins. One assumes the strawberry was chosen because it, too, was a fruit (as the tomato technically is), is red (allowing you to make it from the same material), and was smaller than the tomato, because no one needed a large, heavy sharpener.

It's a plausible theory, and I'll accept it for now, but I observe that the only place I've found it is on sewing websites. No one gives any source for this legend, and the Internet is notorious for having sites that shamelessly copy from each other, without any sort of fact-checking (as the huge number of websites with untrustworthy quotations attests). I haven't been able to corroborate the tomato pincushion story using any non-sewing websites, especially contemporary literature.

How old is the term "pincushion distortion"? Joseph Petzval probably worked out the theory by about 1850, when he designed his portrait lens. Certainly Philipp Ludwig von Seidel had derived the formulae by the 1850s. There have been arguments recently about who deserves primacy for working out the mathematics of the aberrations, since Petzval must have done so to design his lenses, although he published very little regarding this.

But the earliest reference to the phenomenon of distortion I have come across is even earlier than these, in the work of a largely neglected artist named Cornelius Varley (1781–1873). He was the younger brother of watercolorist John Varley, and he was educated by his uncle, a manufacturer of scientific instruments. This unusual coincidence of influences resulted in the younger Varley becoming not only an artist but an expert on optical devices. He made improvements on the camera lucida and the camera obscura, both to be used as aids to drawing. He invented a combination camera lucida and telescope, which he called the graphic telescope. He was awarded the Isis Medal (a medal with an image of the goddess Isis on its obverse and an inscription relating to the achievement on the reverse) from the Royal Society of the Arts for his work, as well as a medal at the Great Exhibition at the Crystal Palace in London. In 1845, he published *A Treatise on Optical Drawing Instruments*,[1] which gives detailed studies and evaluation on optical system design applied to imaging, a perfect book for the emerging field of photography. On pages 7, 12, and 13, he describes "optical distortion," illustrated in his figures 5 and 6, which depict the now-familiar form of a square grid afflicted with what we now term "barrel and pincushion distortion." Varley did not assign them names, just pointing out that they were both sides of the same coin.

The earliest use of the term "pincushion distortion" I can find, using Google Ngram Viewer, is from 1862,[2] with the usage suddenly becoming popular after 1866. That's certainly after the first appearance of square pincushions. Depending upon when, during the Victorian era the tomato pincushions might have originated, it might be before they were around to confuse the matter.[3,4,5,6,7,8]

Notes

1. http://books.google.com/books?id=aLxbAAAAcAAJ&pg=PA3&dq=A+Treatise+on+Optical+Drawing+Instruments,&hl=en&sa=X&ei=pD9WVMzuLc-5uATOvYL4CQ&ved=0CDIQ6AEwAA#v=onepage&q=A%20Treatise%20on%20Optical%20Drawing%20Instruments%2C&f=false

2. George Dawson, "Of Instantaneous Photography," *British Journal of Photography* 9: 138–139. http://books.google.com/books?id=qJwOAAAAQAAJ&pg=PA138&dq=%22pincush ion+distortion%22&hl=en&sa=X&ei=k6NVVOv4Ioy9uASJmYDwAQ&ved=0CCgQ6AE wAA#v=onepage&q=%22pincushion%20distortion%22&f=false

3. http://en.wikipedia.org/wiki/Cornelius_Varley

4. http://en.wikipedia.org/wiki/Optical_aberration

5. http://en.wikipedia.org/wiki/Distortion_(optics)

6. http://www.threadsmagazine.com/item/23383/why-are-pincushions-frequently-made-to-resemble-tomatoes

7. http://en.wikipedia.org/wiki/Pincushion

8. Some websites say that the tomato pincushion originated in Tudor times, with no more references than those who claim they originated in Victorian times (http://www.ehow.com/facts_6767493_history-tomato-pin-cushions.html; http://fromthesehandsblog.blogspot.com/2009/05/pincushion-history.html; http://iroquoisbeadwork.blogspot.com/2012/04/early-beaded-iroquois-and-wabanaki.html). This seems unlikely. Tomatoes were known in Tudor times in Europe, and they were eaten on the Continent, but tomatoes weren't eaten generally until later in England (and in the English colonies). As relatives of the deadly nightshade, they were thought to be poisonous. For the same reason, it seems unlikely that they were regarded as good-luck charms.

21
Preppy Physics

The virtues of Pink and Green—The wearing of pink and green is the surest and quickest way to group identification within the Prep set. There is little room for doubt or confusion when you see these colors together— no one else in his right mind would sport such a chromatically improbable juxtaposition.

—*The Preppy Handbook* (1980)[1]

The *Preppy Handbook* came out in 1980 and revealed to those of us not raised in upper-middle-class WASP households the passwords and clichés of that culture, including the love of the pink and green color combination. I make no apologies for invoking a book over 35 years old. It underwent a rewrite about five years ago. And, as a quick perusal of the Internet shows, the pink and green color combination is with us still.

But I must take issue with the last part of the statement quoted earlier. Although perhaps "no one else in his right mind would sport such a chromatically improbable juxtaposition," Mother Nature does it all the time.

What color is a soap bubble? Soap bubbles are intrinsically colorless. Even if coloring is added to the solution, the film is so thin that it has hardly any effect on light passing through. But, as we learn in our Optics 101 classes and Thin Film seminars, the thin parallel layers of a soap film sport colors due to the effects of interference. Such interference gives colors to not only soap films but also oil films, supernumerary rainbows, Newton's rings, and other phenomena. But the colors these generate are not "all the colors of the rainbow." The rainbow itself, after the initial set of colors (which may not display a full spectrum, depending upon the sizes of the raindrops), turns into a series of bands that can be described as alternating between a sort of aqua green and a magenta pink. Soap bubbles, when they are very thin, may be colorless or blue or yellow, but as they get thicker, they reflect one of two hues—either the aqua green or the magenta pink. The same goes for oil films, Newton's rings, and the color orders of the *isogyre* seen between crossed polarizers.

What we are seeing is, of course, multiple-order interference of a white light source. The ubiquitous pink and green colors are the inevitable result of cascading the interference effects for the many colors making up the white light illuminating source. Since the wavelengths are different, a particular optical thickness of film won't correspond to the same fraction of a wavelength for all colors, and the juxtaposition of them results in white, aqua green, or magenta pink. If the illuminating source were monochromatic, or if we viewed the effect through a narrow-band interference filter, we'd only see alternating bands of bright color and dark bands of destructive interference.

But in nature we don't generally have monochromatic sources—we have broadband white light with extremely short coherence length, so interference effects only show up in a few places where we get interplay between parallel paths of only slightly different length, as in a raindrop, or an ice crystal, or a soap or oil film. In those cases, or the slightly more artificial ones of Newton's rings or crystal polarimeters, we get to see this effect of white light interference colors. Yet this is clearly an important and interesting phenomenon, if one rarely covered in optics classes or texts. It gives a powerful clue to the wave nature of light—as Thomas Young knew. He seized upon these examples as evidence for this alternative to the corpuscular theory.

But people before him saw the colors in supernumerary rainbows, in soap bubbles, or in oil films. What did they make of them? Did these provide clues for the nature and behavior of light? As far as I can tell, they didn't.

An old "Ripley's Believe It or Not" cartoon claimed that Ibn al-Haytham (Alhazen) was drawn into the study of optics by contemplation of the colors of a soap bubble. But I can find no support for this in the traditions surrounding Alhazen. Nor could I find evidence for recognition of this phenomenon until a few hundred years ago. The first reference appears to be from the 13th-century Polish-German Witelo, author of the optics work *Perspectivo*, which describes the colors of the supernumerary bands "as first red then green: and then purple: and again red: and again green: and finally purple." The monk Theodoric of Freiburg described them in the next century. They were remarked on by rainbow investigators Edme Mariotte and Philippe de Lahire, but the most famous description of them was given by Benjamin Langwith (1684–1743), in the *Philosophical Transactions of the Royal Society* in 1722.[2] Its title is pure poetry—"Extracts of Several Letters to the Publisher, from the Reverend Dr. Langwith Rector of Petworth in Sussex, concerning the appearance of Several Arches of Colour Contiguous to the Inner Edge of the Common Rainbow." Langwith was not the first to observe these, but, as Carl B. Boyer observed in his classic work *The Rainbow: From Myth to Mathematics*,[3] it "was cited by others so frequently that their discovery came to be ascribed to him." He describes four observations at different times of rainbows having supernumeraries. One of these lists the colors in each, indexing the bows with Roman numerals. The first bow is the "standard" rainbow:

I. Red, Orange, Yellow, Green, Light Blue, Deep Blue, Purple
II. Light Green, Dark Green, Purple
III. Green, Purple
IV. Green, Faint Vanishing Purple

The colors in a soap bubble were remarked upon by Robert Boyle in his *Experiments and considerations touching colours first occasionally written, among some other essays to a friend, and now suffer'd to come abroad as the beginning of an experimental history of colours* (1664),[4] Experiment XIX. The interference colors seen between dielectric surfaces separated by a thin layer of air was first remarked on by Robert Hooke in his classic *Micrographia* (1665) in chapter IX—"Of the Colours Observable in Muscovy-Glass, and Other Thin Bodies." "Muscovy-glass" is mica, which forms thin parallel sheets, which mat become slightly separated at one end, while still connected at the other. Hooke noted the colors and made a clear connection with supernumerary bows:

The consecution of those Colours from the middle of the spot outward being Blew, Purple, Scarlet, Yellow, Green; Blew, Purple, Scarlet, and so onwards, sometimes half a score times repeated, that is, there appeared six, seven, eight, nine or ten several coloured rings or lines, each incircling the other, in the same manner as I have often seen a very *vivid Rainbow* to have four or five several Rings of Colours, that is, accounting all the Gradations between Red and Blew for one.

The phenomenon is basically the same as that seen by Isaac Newton between a flat glass plate and a portion of a glass sphere, and now called "Newton's rings." Newton himself gives a rather lengthy description of how the different groups of colors, or Orders, gradually change from a full spectrum to a degraded set of colors.[5]

Newton's observation that the colors in this experiment came in repeating sets might have suggested to him the idea that colors operated in a way analogous to musical tones, which also are grouped together in such "Orders," which we call "Octaves." Newton carried the analogy further by calculating the lengths associated with the colors, which he obtained from the calculated "sag" between the spherical surface and the plane glass. He then showed that these lengths were proportional to the lengths of the strings of an instrument producing the characteristic tones of our musical scale— *do, re, me, fa, sol, la, ti*, and then *do* again, one octave up. To another mind, this might have been an irresistible analogy to wave motion, since the plucked strings clearly sustained wave motion in characteristic wave lengths. But Newton was evidently able to resist the notion, being committed to the corpuscular theory of light. Nevertheless, when Thomas Young finally was able to perform the first measurements of the wavelengths of light, using a finely engraved instrument scale as a diffraction grating, it was Newton's characteristic lengths that he used for his comparisons.

For years afterward, anyone discussing what we would now call the different orders of white light interference called them "Newton's Scale of Colours." Ever since Newton demonstrated that white light was actually composed of the different colors, it had been clear that such white light interference effects as the alternating aqua and pink bands were simply manifestations of this, although no one seemed to offer any suggestion or calculation regarding why the juxtaposition of different orders of colors resulted in only those colors and white, rather than, say, yellow, or orange, or deep blue. In fact, there was not yet a developed theory and science of color vision yet established with which to make such an explanation. The best that could be done was to render the appearance of such interference and to see what use could be made of it.

One of the more important uses appeared in 1888 in the book *Les Minéraux des Roches* (The Minerals in Rocks) by August Michel-Lévy, inspector-general of mines in France, and director of the Geological Survey (and coauthored by Antoine François Alfred Lacroix). Included in the volume was a large chart showing white light interference varying with thickness, which was folded vertically in two and horizontally in four. Running from the lower left corner of the chart are a series of straight lines corresponding to different values of birefringence (which are labeled at the other end of each radiating line). Knowing two of the three values of sample thickness, birefringence, and interference color (and order), you can deduce the third. Michel-Lévy thoughtfully included the names of common minerals corresponding to each of the lines of constant birefringence so that the chart could be used to identify minerals.[6]

But Michel-Lévy obtained his chart by comparison with known samples—it was not derived from first principles. For that, the theory of color was needed. Many people contributed to the development of color theory that culminated in the 1931 CIE Color Space, but the chief contributors were probably William David Wright and his collaborators at Imperial College in London in 1927–1929 and John Guild of the National Physical Laboratory in London,[7] who independently studied human vision and response to color, producing the prototypes of what would become the CIE Chromaticity Diagram. In 1931, the *Commission Internationale de l'éclairage* held its eighth session in Cambridge, England, and formalized its Color Space, the Standard Observer, the Color Matching Functions, and the Standard Illuminants A, B, and C. This provided the framework for any calculations of observed color from a knowledge of the emission or transmission spectrum of the material and a knowledge of illumination sources. The basics were set down, for instance, in such book as the *Handbook of Colorimetry*,[8] issued in 1936 by "The Staff of the Color Measurement Laboratory, Massachusetts Institute of Technology, under the direction of Arthur C. Hardy."

So, with knowledge of the interference transmission of a thin film as a function of thickness, wavelength, and refractive index (also as a function of wavelength), and armed with the new theory of colors, it should have been a simple matter to calculate the colors of thin film white light interference from first principles, thus confirming the work of Michel-Lévy and others, and closing the circle of theory and experiment regarding those colors.

Except that, for some reason, it didn't happen, for quite a while.

Exactly why this calculation was not made isn't clear. Perhaps it was thought to be too simple an exercise, like the parallel exercise of using Rayleigh's law of scattering to derive the CIE coordinates of the blue sky. As far as I know, this was not the subject of any paper in the early days of color theory, either.

I know that, after being assigned the "blue sky" calculation in a graduate school course, I went on to calculate what the color of a thin film was as a function of optical thickness, and plotting the trajectory on the CIE 1931 chart. Not that I was the only one to do so. In the "Amateur Scientist" column in *Scientific American* for September 1978, physicist/columnist Jearl D. Walker describes the calculations of Benjamin Bayman and Bruce G. Eaton of the University of Minnesota, who did exactly the same thing.[9] Their paper was unpublished, and as far as I can tell, it remains so, but at least Walker's column preserves the CIE trajectories they calculated. A similar series of calculations were made by Dietrich Zawischa of Leibnitz University in Hanover, Germany.[10] In recent years there have been several papers on rendering soap bubble colors as an exercise in computer color modeling.[11]

About the only work prior to 1960 that I have found on the subject is that of Hiroshi Kubota of the Department of Applied Physics at Tokyo University in the early 1950s.[12] But Kubota did not take the calculation to the point of plotting the trajectory between aqua and pink colors, as these other calculations do—his represents the "rainbow" portion of Newton's rings.

And yet recent papers indicate that there is value in performing these calculations and comparing them against the Michel-Lévy chart. Bjørn Eske Sørensen's *A revised Michel-Lévy interference colour chart based on first-principles calculations*[13] calculates the color chart "from scratch" and shows that some previous versions improperly represent the widths of colors and their shades. Clearly, there's value in performing these calculations, even if they only seem like a student exercise.

Notes

1. Lisa Birnbach, Jonathan Roberts, Carol McD Wallace, and Mason Wiley, eds., *The Preppy Handbook* (New York: Workman Publishing, 1980).
2. Benjamin Langwith, "Extracts of Several Letters to the Publisher, from the Reverend Dr. Langwith Rector of Petworth in Sussex, concerning the appearance of Several Arches of Colour Contiguous to the Inner Edge of the Common Rainbow," *Philosophical Transactions of the Royal Society* 32 (January 12, 1722): 241–245.
3. Carl B. Boyer, *The Rainbow: From Myth to Mathematics* (Princeton, NJ: Princeton University Press, 1987), 277.
4. http://quod.lib.umich.edu/e/eebo/A28975.0001.001/1:6.4.19?rgn=div3;view=fulltext Hooke's *Micrographia* is available on the Internet in several places, most notably Project Gutenberg here: http://www.gutenberg.org/ebooks/15491
5. It is interesting that Newton's rainbow colors of Orange and Indigo make no appearance here. This is perhaps not because he didn't see them, but because this description was written in the early stages of the book, when he still spoke of *five* colors in the rainbow. Later, he made the analogy between the tones of a musical scale, in which there are seven distinct tones in an octave before repeating the base tone—*do, re, me, fa, sol, la,* and *ti* before returning to *do*. The addition of two more "tones" required the addition of two more colors—Orange and Indigo (Langwith's "Light Blue").
6. Use of the chart here: https://www.mccrone.com/mm/the-michel-levy-interference-color-chart-microscopys-magical-color-key/ On Michel- Lévy, see here: https://en.wikipedia.org/wiki/Auguste_Michel-L%C3%A9vy
7. William David Wright, "A Re-determination of the Trichromatic Coefficients of the Spectral Colours," *Transactions of the Optical Society* 30, no. 4 (1929): 141.
8. http://science.sciencemag.org/content/85/2214/545
9. http://optica.machorro.net/Optica/SciAm/Bubbles2/1978-09-body.html
10. "How to Calculate and Render Colours—Thin Films as an Example," circa 2005. https://www.itp.uni-hannover.de/~zawischa/ITP/soapfilmcalc.pdf
11. See, for instance, David Harju and Simon Que and references therein—https://graphics.stanford.edu/wikis/cs348b-08/David_Harju's_and_Simon_Que's_Final_Project_Writeup—although computer modeling incorporating thin film interference colors goes back much earlier than this.
12. Hiroshi Kubota, "On the Interference Colors of Thin Layers on Glass Surface," *Journal of the Optical Society of America* 40, no. 3 (March 1950): 146–149.
13. Bjørn Eske Sørensen, "A Revised Michel-Lévy Interference Colour Chart Based on First-Principles Calculations," *European Journal of Mineralogy* 25, no. 1 (2013): 5–10. See also Bjørn Eske Sørensen, "A Revised Michel Lévy Interference Color Chart Based on First Principles Calculations," *Geological Society of America Abstracts with Programs* 44, no. 7 (2014): 5–10.

22
The Sandbow

In the morning there was khamsin [a sandstorm]and we saw a sandbow. It was on a level with the sun, and not opposite it as in rainbows, but about 30° from it; not the shape of a rainbow, but of a nebula; all the colours perfect. It had a most singular effect; it was about midday, so that the top of the pillar of sand must have reached that height.
—Florence Nightingale, Claydon Diary, December 29, 1849

This opening quote records an observation by Florence Nightingale—yes, *that* Florence Nightingale, the founder of modern nursing—while she was on a boat trip down the Nile River, at the island of Metareh, north of Thebes. Within a few days of this, she believed she received a "call to God" that resulted in her devoting her life to nursing. It's one of only a handful of references I have found to the phenomenon of the sandbow, which has fascinated me for some time.

My interest was stirred by considering what other materials, besides water drops and ice crystals, might be responsible for "bows" and arcs analogous to rainbows and ice crystal arcs and haloes. There are cases of haloes around the sun or moon being caused by grains of pollen (which can cause elliptical haloes) or by droplets of plant oils after forest fires or by grains of volcanic dust (the "Bishop's Ring," named after Reverend Sereno Edward Bishop of Honolulu, who, though a reverend, was not actually a bishop). But, as far as I am aware, there have been no reports of any meteorological optical phenomena besides haloes due to these causes—no rainbows or arcs that require a more complex path through the material.

There ought to exist odd ice crystal arcs and "rainbows" from exotic liquids and ices on other planets and moons, where methane ice is a common material, and calculations regarding these have been made, but these unusual optical phenomena do not occur on earth. The requirements for an appropriate material are that it should have very clean surfaces and a well-defined and characteristic form. Spherical raindrops and hexagonal ice crystals are ideal for there, and droplets of plant oils or pollen grains would work as well. But what other materials might work? Most other liquids aren't really common in bulk on the earth's surface. Crystals tend not to be in replicated small units, with identical angles. When ground, they tend to be irregular in form and with dirty surfaces.

It occurred to me that salt crystals, ordinary NaCl, such as one might find in great abundance on the salt flats of Utah, might fit the bill. Among newly formed salt grains from evaporation of the Salt Lake there might be plentiful tiny crystals with clean and open faces.

Alas, NaCl crystals won't work. They have 90° edge angles, just like those that contribute to the 46° ice crystal halo, but this angle is too large for a salt halo or arc. The condition for such an angle to work as the roof angle of a minimum-deviation prism is that the refractive index of the material be less than $\sqrt{2} \approx 1.414$, and that's a pretty severe restriction. Ice works, because its refractive index (at the sodium D line) is a very low 1.309. But NaCl has an index of 1.516—any light entering would be totally internally reflected. In fact, all those cubic alkali halide crystals, like the common KCl ($n_D = 1.4902$), suffer from the same problem. Only the alkali fluorides have refractive indices low enough to qualify. But they have a different problem. Leave a piece of potassium fluoride sitting out on a desk in humid weather and you would come back later to find a puddle of salty water. Like the other fluorides, it's very hygroscopic; it will first attract water from the air that will cover its surface and then it will dissolve in it. Only lithium fluoride is free of this problem. Light, hard, and tough, LiF is transparent farther into the ultraviolet than any other solid. Its index of 1.3915 makes it a perfect candidate for this—except that LiF is virtually never found in nature (except, perhaps, as a chance inclusion in other lithium minerals). It is an artificial product, created by chemically reacting hydrofluoric acid or other source of fluorine with lithium compounds.

But while looking into the possibilities of crystals from Utah's Great Salt Lake as the bases for alternative arcs and rainbows, I stumbled onto an unexpected and intriguing possibility—sand. Sand doesn't really seem like a viable alternative. Common sand is mainly silicon dioxide, which occurs in a number of crystalline forms with a variety of refractive indices, most of them greater than 2.5. The various crystals have facial angles that might allow for interesting prismatic effects, but sand grains are rarely crystalline. They are generally broken off and irregular, or rounded by abrasive action. One wouldn't expect them to have the uniformity required to produce a bow or arc.

In the June 21, 1901, issue of *Science* appeared a letter entitled "A Sandbow—An Unusual Optical Phenomenon." Its author was James E. Talmadge, professor of geology at the University of Utah (where he had served four years as the president) and future member of the Quorum of Twelve Apostles of the Church of Jesus Christ of Latter Day Saints. In the letter he related that he had been on Antelope Island—the largest island in the Salt Lake—on May 16 and had seen an unusual phenomenon. As he looked eastward from the eastern slope toward the mainland, he and his party saw a bright rainbow having twice the width of a normal bow, yet there was no rain present. The sand, he observed, was "oolitic sand," which is unusually spherical in shape and composed of calcareous material, with very uniform size and having a polished, pearly luster. A wind blowing toward the mainland had lifted large quantities of the sand into the air, and this, he believed, was responsible for what they saw. The "sandbow" was about 40° from the antisolar point (about where a rainbow ought to be), with red on the outside. There was a faint secondary sandbow beyond this, with the colors in reverse order (this exactly matches the color orders in the primary and secondary rainbows).

Talmadge had no explanation for the observed effect: "The production of a color bow by reflection from the outer surfaces of opaque spherules is a new phenomenon to the writer. It is inexplicable on the principle of refraction and total reflection from the interior of transparent spheroids, according to which the rainbow is generally explained." At the beginning of the piece he had stated that "The following description, based on personal observation, is presented without discussion of the optical principles involved." Undoubtedly, this was because he couldn't come up with a convincing explanation. Good scientist that he was, though, he reported what he had seen as accurately as possible. He concluded: "If phenomena similar or analogous to the foregoing have been observed, reports of the same would doubtless be of instructive interest." They certainly would be. I have looked for similar reports of sandbows and found only two others—neither obviously involving oolitic sand. One is Florence Nightingale's report cited earlier. She doesn't state that oolitic sand was involved, although it is present in Egypt. The other is recounted in a book about divine revelation, in which a rainbow observed made of dry dust, rather than rain, is interpreted as a sign from God. In addition to these, I have found three fictional accounts of sand or dust rainbows.

In looking over these, I find that I cannot rely on the fictional cases. Ms. Nightingale's description doesn't sound like a rainbow at all. In fact, it is a perfect description of a sun dog or parhelion, an effect caused by light being refracted at the angle of minimum deviation by hexagonal (or trigonal) ice crystals suspended with their broad surface horizontal in the air. It's actually a very common phenomenon—about an order of magnitude more common than a simple rainbow, but one rarely noticed by most people. Contrary to expectation, the ice crystals can form even in the sky over a desert, at sufficient altitudes. It appears that Ms. Nightingale's reason for thinking this effect was due to sand was because no rain was present, but sun dogs are generally seen when no ice is in the vicinity of the ground.

Similarly, rainbows can be formed when there is no rain in the observer's immediate vicinity. Regarding Talmadge's particular case, he says that the oolitic sand was about the size of number 8 to number 10 shot. That would make them somewhere around 2 mm in diameter. Oolitic sand is due to some nucleus (often a brine shrimp fecal pellet in Salt Lake ooids) that has layers of calcium carbonate ($CaCO_3$) precipitate out on it, forming concentric layers. Modern ooids are less than 2 mm in diameter (matching Talmadge's description). Calcium carbonate has two crystalline forms, both of which have been observed in ooids—calcite and aragonite. Aragonite is typical of Salt Lake ooids. At 2.83 g/cm^3, a 2 mm diameter ooid would weigh 12 milligrams and could be raised by winds. It's also opaque.

I entertained ideas of thin films on the individual grains, or of light passing through one layer of the concentric shells. I could imagine interference between the reflections from one such grain causing an angle-dependent rainbow, but it was hard to see how a vast collection of such grains could have the uniformity of thickness required to give the same effect, and it was hard to understand the cause of a secondary, *reversed* rainbow by this mechanism. (Aragonite is a biaxial crystals, by the way, and calcite is uniaxial. The optics of reflection from such a layer would be very complex.)

The fact that the observed "sandbow" was in the same location—give or take a couple of degrees—as an ordinary rainbow, and that the orders of colors in both the primary and secondary "sandbows" were identical to those of an ordinary rainbow makes it highly probable that Talmadge was, indeed, seeing an ordinary rainbow. Because of the lack of observed rain and the extraordinary width, he was convinced that he was seeing something unusual. But the raindrops need not have been nearby at all, but suspended in the air above (I can attest from my own experience that rainbows are very often seen against the Wasatch Mountains to the east of the Salt Lake—clouds often drop their moisture, often pulled from the lake itself, when they encounter the temperature change at the mountains), and the widths of rainbows do vary with the sizes of the drops. In the more than a century since this observation, there appears to have been no other observation of such an effect, and it seems likely that Talmadge was simply mistaken.

References

Howell Kopp Heffern, Jean. *Finding God's Fingerprints: A Daily Devotional*. Bloomington, IN: Authorhouse, 2004.

McMurtree, Larry. *The Berrybender Narratives*. New York: Simon and Schuster, 2011. p. 503.

Shah, Lisa Meredith. *Escape to Ecclesia*. Bloomington, IN: Balboa Press, 2012.

Silberstein, Stephanie. *Winter's Silence*. Dunn NC: Narrow Path Publishing, 2008. p. 15.

Talmade, James E. "A Sandbow—An Unusual Optical Phenomenon." *Science* (New Series) "A Sandbow—An Unusual Optical Phenomenon" 13, no. 338 (June 21, 1901), 992.

Vallée, Gérard, ed. *Florence Nightingale on Mysticism and Eastern Religions*. Waterloo, Ontario: Wilfrid Laurier University Press, 2003.

23
Mistbow versus Glory

Recently I had a discussion on a web board that I frequent. Someone had asked what things were like before the creation of the Rainbow in the book of Genesis—how could you not have a rainbow in a world consistent with our physics? I suggested a number of ways that it would be possible for the rainbow not to be visible—ways that wouldn't require bending the laws of physics too seriously. These ranged from the sort-of-obvious (perhaps everyone was color blind back then) to the bizarre (was it possible that raindrops weren't spherical until after the Flood?). One of the more interesting, I thought, was my suggestion that, until then, water drops only came in minuscule sizes, so that one would only see mistbows, otherwise known as fogbows. When the sizes of drops are on the order of microns, the colors are completely absent, and you only have a white bow. After the Flood, I suggested, God let the droplets get bigger, and so full rainbows, with all their colors, would form.

But, someone replied, that wouldn't completely remove color from atmospheric optics, because you'd still have glories. That's true, I replied, but the viewing conditions for a glory are completely different. You still wouldn't have colored rainbows.

But the interchange made me stop and think. Why *were* glories colored and mistbows white? Both were the result of very small drops, typically tens of microns in diameter. Both were described mathematically by the theory of Mie scattering. Yet glories are very brightly colored and mistbows are not. Everyone is familiar with rainbows. A mistbow or fogbow is simply the same thing with smaller drops, and it appears as a white rainbow, without colors. A glory is a less familiar phenomenon, in which a very tight multicolored circle is observed around the shadow of the observer when looking at a cloud or fogbank. It used to be that you had to be atop a mountain or looking into a mist-filled valley to see one. Today, however, they are very commonly seen by airplane passengers, looking down at the shadow of the plane on a cloud (Figure 23.1).

If geometric optics held in all cases (if, that is, the wavelengths of light were infinitesimally small, beyond measuring), then all rainbows would look the same. But when the sizes of the drops begin to approach the scale of the wavelength of light, then diffraction starts to become increasingly important. The tendency of raindrops to act like prisms, dispersing the different colors, is counterbalanced by the diffraction of light through that raindrop, which also acts somewhat like an aperture. The smaller the drop gets, the broader the angular extent of the emerging light, until for sufficiently small droplets the colors completely merge together. As a result, the rainbow is completely white. The appearance of the rainbow varies with drop size even before the colors merge, and you can tell the size of the drops responsible for the bow by the number and distribution of the colors.

Much the same thing happens with ice crystals. The phenomenon called "sun dogs" or "parhelia" or "mock suns" is due to an extremum in the light refracted from

Figure 23.1 Glory seen on a cloud from an airplane. Notice how the glory is centered on the airplane's shadow.
Shutterstock

hexagonal ice crystals that are oriented in one direction. They are actually portions of the 22° halo that lie horizontally on either side of the sun. (If you hold your arm outstretched, with the fingers of your hand spread wide, then 22° is the distance between the tips of your thumb and little finger.) Sun dogs are actually very common—about 10 times more common than rainbows—but are rarely noticed. This is partly because people are unfamiliar with them, but also because most sun dogs are not very impressive. When they are created by large ice crystals, then the full spectral magnificence of the effect can be seen. The sun dog looks like two bright spots, one on either side of the sun, brilliant red on the inside going toward bright blue away from the sun. Most of the time, however, the effect is due to tiny crystals, in the form of cirrus clouds, and the diffraction from passing through such small crystals undoes the color separation, so the sun dogs appear as merely bright spots in the clouds or as an unremarkable brown smudge on the sky (Figures 23.2, 23.3).

Not so with glories, however, which are always brightly colored. Yet they result from light that also passes through similar-sized droplets. In fact, there are photographs of glories and fogbows observed simultaneously, resulting from the same cloud bank. That the light passes through the drop is proven by the fact that the angular size of the glory depends upon the size of the droplet. So why are glories colored? Shouldn't the same diffraction that removes the colors from mistbows and sun dogs have the same effect on glories?

That both phenomena can be described by Mie theory clouds the situation somewhat. We can take a simpler, approximate model to gain greater insight into the

Figure 23.2 Sun dogs visible on either side of the sun are not very spectacular, with little color. That is because these sun dogs are due to very small ice crystals, and diffraction due to the light passing through such small crystals undoes the prismatic color separation.
Shutterstock

Figure 23.3 A sun dog due to a large ice crystal shows bright colors, because diffraction from a large crystal doesn't significantly affect the width of each color.
Shutterstock

physics involved. Imagine treating these, instead, as two-dimensional problems. This is what George Biddell Airy did in the early 19th century when he derived the physical optics explanation of the rainbow. The rainbow, as an interference phenomenon, is caused by light near the extremum of the plot of diffraction angle versus "impact parameter" (the fractional radial distance from the center at which the light enters the spherical raindrop). Because there is a critical angle at which the deviation is a maximum, on either side of this maximum are two values for which the light emerges at the same angle, but with a small path difference. This, as Thomas Young realized at the beginning of the 19th century, meant that there could be interference. Airy's genius was to realize that this interference was due to a cusp in a wavefront that could be described as a cubic function of angle, and to use that to calculate the effect of the full wave interfering with itself, treating it as a series of Huygens wavelets. The result was the Airy Rainbow Integral, which gives the light distribution to very good approximation and accounts for effects like the supernumerary bands.

The glory, on the other hand, is rather more complex, and its theory is still not standing on a satisfactory theoretical footing. Essentially, light enters the drop near one side, is internally reflected from the rear surface, then exits from the other side, counterpropagating along the reverse of the incident angle. The problem is that the angles do not work out for water, with its reflective index of 1.33. In order for light to be perfectly retroflected, the index must be equal to or greater than the square root of two (1.414). To make the angles work out right, theoreticians have had to take recourse in the light traveling a short distance along the surface of the drop due to surface waves. Everyone is very apologetic about this explanation and assures the student that it's all very proper. But you can tell they're embarrassed about it.

Nevertheless, the model does work. The glory is the result of interference between light that travels one direction (clockwise, say) around the drop with another ray that travels counterclockwise about the same path. In two dimensions, then, the glory results from interference between two points at opposite ends of the drop. To within a factor of two, glories look like two-slit interference where the slit separation is the width of the drop.

After all that, though, the diffractive spread of light rays from the drop in the case of a glory is about the same as the diffractive spread from the same drop for a mistbow. So why is one colorless and the other colorful?

The answer is that the rainbow or mistbow must always be near the geometric rainbow angle. There's a slight variation with wavelength, but the extrema for all the colors span only about a degree, from 41° from the antisolar point to about 42°. The maximum of the color as determined by the airy integral is only slightly shifted inward from this, regardless of the size of the drop, with the result that for large drops the colors are separated, because the widths of the different colors are narrow, and the small offset for each color's location lets you see each one independently. But as the drops get smaller, the locations of each color remain in the same place, but the widths of each color band increase, with the result that they overlap (Figures 23.4, 23.5).

For the glory, however, the pattern is "pegged" at the antisolar point, in the center of the observer's shadow, and both the pattern size and the width of each color depend upon the droplet size. As the droplet gets smaller, the pattern increases in size at the same rate that the locations of the maxima increase, with the result that the maxima

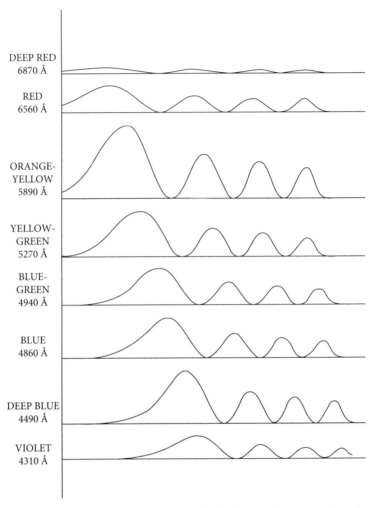

Figure 23.4 Intensities versus angular position for different colors in a rainbow for a large raindrop. Notice that the leftmost main band shifts slightly with each color. When the drop is large, the colors don't overlap very much. As drop size decreases, the main bands stay in the same locations, but they get wider, eventually completely overlapping so that the color separation is lost.

From R. A. R. Tricker, *Introduction to Meteorological Optics*.

don't overlap. The glory formed by smaller drops looks like a larger version of the glory formed by large drops, with the same color banding (Figure 23.6).

Incidentally, although the glory is brightly colored, it is not colored with "all the colors of the rainbow." The glory does not give a spectrum, but has the characteristic look of white light multiorder interference, with a dark center surrounded by a white ring, then a yellow, a red, and a blue. This is followed by those characteristic pink and green bands that are seen in so many cases of white light interference. The colors in a glory more closely resemble those in an oil film or a soap bubble that those of a rainbow.[1]

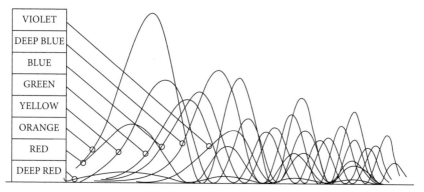

Figure 23.5 Intensities vs. angular position for different colors in a rainbow for a large raindrop, plotted on the same graph. The red band of the rainbow is on the left, and the leftmost peaks of the different colors are somewhat separated and distinct, but the subsequent peaks overlap and jumble together.
From R. A. R. Tricker, *Introduction to Meteorological Optics.*

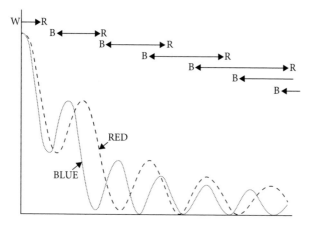

Figure 23.6 Intensities versus angular position for different colors in a glory for a large raindrop. As the drop changes size, the first nonzero peak not only widens but also shifts position away from the center. As the drop gets smaller, the peaks widen, but they also get farther apart, so the colors do not overlap as in a rainbow.
From R. A. R. Tricker, *Introduction to Meteorological Optics.*

Note

1. "Preppy Physics," *Optics & Photonics News* 30, no. 2 (February 2019): 22–24 or Chapter 21 in the present volume.

24
Edible Optics

Some time ago, I wrote about edible lasers.[1] This time I'm looking at something less ambitious—edible optics. It occurred to me that the "candy glass" used in movie-making in place of real glass is both highly transparent and edible. Are there other materials that could be used to construct edible optical elements?

For purposes of this discussion, I am limiting it to solid materials with some stability, and ideally not saturated with water. Water and water solutions are transparent, of course, and you can vary the refractive index by raising the concentration of salt or sugar in water. Alcohol, too, is potable. But these don't give you a lot of range, and they have to be contained by something solid, which will necessarily dissolve in water, if it is to be edible. Also, a lot of substances owe their transparency precisely to the presence of copious amounts of water—gelatin, "cellophane noodles," and so forth. And, although many students have made lenses out of ice, I'm restricting this to room temperature materials. What can you use to construct an optical device in the visible with solid, dry materials that you can safely eat?

One edible material that has long been used to construct optical elements is sodium chloride. Starting in the mid-19th century, this was one of the few high-quality materials available for constructing lenses for the infrared. Spectroscopes used "rock salt" prisms for covering broader wavelength ranges than glass prisms would allow, and salt windows are still widely used. One problem with salt is that it is slightly deliquescent and is attacked by humid air. Rock salt prisms used to be stored when not in use in desiccator jars, or they were attached to electric heaters to drive off moisture. One experiment I worked on with a large salt window required us to seal it with desiccant and place a heat lamp over it when not in use. Its refractive index is 1.54 at the sodium D line, with a V number of 42.9 and a dispersion of $-0.0647 \ \mu m^{-1}$.

Potassium chloride is also edible, being used as a common "salt substitute," and it, too, can be used as an optical material. Its refractive index is 1.49 at the sodium D line, with a V number of 44.4 and a dispersion of $-0.05625 \ \mu m^{-1}$.

Rochelle salt (sodium potassium tartrate) is easily grown into large crystals. It is also deliquescent and has a refractive index of 1.59. It is used as a food additive and laxative.[2]

As I mentioned earlier, sugar can be made clear and formed into sheets and cast in the form of lenses. The usual formula for "candy glass" consists of sugar, cream of tartar (to prevent crystallization), and corn syrup. Properly done, this yields a good clear sheet, but different mixtures can come out yellow and very flexible.[3] Worse, such sugar mixtures tend to be deliquescent and can "melt" out of shape.

A solution that avoids the problems of deliquescence, yellowing, and softening is to use single-crystal sugar. The coarsest size of commercial single-grain sugar is called "coarse white sugar" or "decorating sugar" and typically measures no more than 1–2 mm on a side. But larger crystals are certainly possible. Everyone is familiar with polycrystalline "rock candy." These are polycrystalline, with individual crystals generally measuring a few millimeters on a side, but larger crystals of 10–20 mm can be seen in some samples of rock candy. An Internet site has pictures of single crystals measuring over 20 mm.[4] Technical papers about the growth of larger crystals have been written.[5] The largest crystal I have uncovered is a 1941 report of a 7 lb single crystal, which took 14 years to grow from solution.[6] Unfortunately, only the center 1 inch or so was perfectly optically clear. The refractive index of sucrose is 1.57.

A faster way to produce something similar, with great clarity and large size, is to use Isomaltitol, usually called Isomalt.[7] This is a polyol, or sugar alcohol, with the chemical formula $C_{12}H_{24}O_{11}$, and the daunting IUPAC (International Union of Pure and Applied Chemistry) name of (2R,3R,4R,5R)-6-[[(2S,3R,4S,5S,6R)- 3,4,5-trihydroxy-6-(hydroxymethyl)- 2-tetrahydropyranyl]oxy]hexane- 1,2,3,4,5-pentol. First produced in the early 1980s by the German company BENEO-Palatinit, it is also a component of the sweetener DiabetiSweet. It was approved for use in the United States only in 1990, and it is now used to produce decorations for high-end cakes. Isomalt can be melted in the microwave, cast in silicone molds, and hardened in 15 minutes. Commercially available forms include "diamonds,"[8] but there's no reason that plano lenses could not be made in silicone molds or more complex forms from two-piece molds. Indeed, Isomalt has already been molded into lens forms by cake decorators making the forms of cameras on cakes, although these lenses were not intentionally functional.[9]

Isomalt is very hard, not deliquescent, and resistant to moisture. In liquid form, Isomalt has a refractive index between 1.4791 and 1.4816, and the solid likely has nearly the same value.

One could make a magnifying glass of any of these, or a telescope or microscope from combinations of lenses. The idea of a sweet-and-salty achromat is appealing, but needs further work.

There are other edible materials that transmit light, but they tend to be translucent and/or must be very thin. I thereby discount clear edible thin films and most other somewhat clear materials.

The aforementioned materials can be used to fabricate lenses and prisms by grinding and polishing or, in the case of Isomalt, by careful casting. But there's another optical device that can be fabricated from sugar—optical fibers. Sugar can easily be spun into long fibers, familiar as cotton candy or floss. US Patent #6,416,800, granted to Paul J. Weber and Brian D. Andresen, is for an edible fiber-optic display made of such sugar fibers, illuminated at one end by a conventional light source. The sugar fibers need not be the end of the optical chain, they explain, but may be used to convey the light to other edible structures that can then light up, including candies, frozen confections, and gelatin ones.

One of the inspirations for the patent (the patent cites it) is an article by Cynthia Graber in the magazine *Scientific American Explorations* entitled "Make Edible Fiber Optics," which describes making fiber-optic conduits out of clear gelatin.[10] Gelatin, of course, contains water, so this particular use goes against my earlier proscription, but it is a pretty simple home lab experiment for kids, and it is less messy than "fiber-optic" experiments using water streams and the like.

The use of sugar fibers as an analog to glass fibers actually played a role in developing glass fiber-drawing technology. According to science writer Jeff Hecht, trying to work out the mathematics to model the simultaneous drawing of a core and cladding from a double crucible was so taxing that in 1968 researcher Richard Dyott of the British Post Office Research Station in Dollis Hill, London (which was a major center for fiber-optic research in the 1960s) tested the process for real, using sugar instead of glass. Molten sugar formed a thick liquid from which thin fibers could be drawn, just like glass, but it melted at 107°C (225°F), much lower than the melting point of glass, and was much less corrosive as well. This let him experiment at lower temperatures and with brass crucibles instead of platinum. Later, George Newns continued the work, substituting a molasses "core" and a clear sugar "cladding" for the differently dyed sugar Dyott had used. The resulting fiber was useless as a means of conveying light, but it served as an analog for modeling the processes, and a much more accessible one than the recalcitrant mathematical model.[11]

Yet another possibility violates my original structure against compositions that are mostly water. In 2014, the Japanese confectionary store *Kinseiken Seika* came out with a treat called *mizu shingen mochi*. *Shingen mochi* is a rice cake snack. *Mizu* means simply "water." *Mizu shingen mochi* is thus "water-rice-cake," but the name is usually translated as "water drop cake" or "raindrop cake." It has agar and spring water added to the base, and the result looks like an overlarge drop of water, very transparent. It's served with a brownish sweet sugar topping (which rolls off and pools at the base) and a powdery "crumble" of soya bean powder. If not eaten in half an hour, it dissolves. It would make a good, if nonspherical lens. Placed atop a picture, it would probably make a good magnifier, for as long as it lasted.[12]

I'd just like to finish with the caution—if it's necessary—that even though you *can* eat these materials, it's not a good idea to do so in quantity. It's not good to overindulge in salts (or even sugar, for that matter), and eating a large amount will give you gastric distress. And I observe that Rochelle salt and Isomalt are very effective laxatives as well.

To my surprise, after an abbreviated version of this appeared in *Optics & Photonics News*, someone has proposed practical applications for edible optics and has constructed some devices. Takahiro Uji, Yiting Zhang, and Hiromasa Oku, all of Gunma University in Kiryu, Gunma, Japan, delivered a talk and a paper on "Edible Retroreflectors" recently.[13]

The form might seem strange, but they propose the use of retroreflectors as markers for augmented reality (AR) and mixed reality (MR) for circumstances where there are low light levels, and external lighting must be supplied and easily identified. Edible AR markers and QR codes have already been fabricated, so why not retroreflectors to mark positions?

They decided to use a collection of corner cube reflectors molded in an agar mixture that is similar to *mizu shingen mochi*. This was the most highly transparent material they tested (most edible materials have a somewhat yellowish tinge). They added a lot of sugar to the mix to increase the refractive index. Mochi is largely water, and this ought to have a refractive index close to 1.33. Oku et al. report that their mixture has a refractive index 1.4335. It is sufficient, when the retroreflector is in air, to allow the total internal reflection necessary for the retroreflector to work. The resulting retroreflector array provided less reflection than a standard retroreflector, but enough to

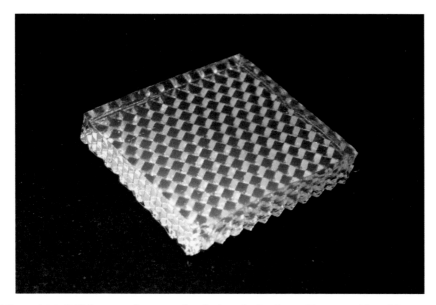

Figure 24.1 Edible retroreflector made of gelatin by Professor Hiromasa Oku of Gunma University, Maebashi, Japan.
Courtesy of Professor Hiromasa Oku, Gunma University.

allow their machine vision system to lock onto it and follow the reflector.[14] The same team has also produced a lens made of agar[15] and a retroreflector of sugar (Figure 24.1).[16]

It seems to me that they could also get good results using isomaltose, which retains its shape better and doesn't dry out so easily. In addition, there are alternative shapes (although the corner cube is probably the most widely used). One that comes to mind is the use of small spheres to create a *heiligenschein* effect, as with glass spheres dusted onto a light-colored background. The spheres need not even be precisely round, since irregular shapes can produce retroreflection through the sylvanshine effect. You can create acceptably round spheres with clean surfaces by dropping the material into a dense supporting medium like oil.[17]

Notes

1. Stephen R. Wilk, "Edible Lasers: What's the Next Course?" *Optics & Photonics News* 20, no. 5 (May 2009): 14–15. Also chapter 27 in Stephen R. Wilk, *How the Ray Gun Got Its Zap!* (New York: Oxford University Press, 2013).

2. http://www.homepages.ucl.ac.uk/~ucfbanf/general/crystal.htm. See also N. A. Romanyuk and A. M. Kostetskii, "Dispersion and Temperature Dependences of Refractive Indices of Rochelle Salt Crystals," *Soviet Physics-Solid State* 18 (1976): 867–869.

3. This site tells you how to make a candy glass magnifier: http://www.instructables.com/ id/Cook-up-an-Edible-Magnifying-Glass/; here are other sites: http://www.csiro.

au/helix/sciencemail/activities/SugarGlass.html; http://www.lehigh.edu/imi/pdf/CandyGlassRecipe.pdf. I add that "candy glass" manufacturers for movie props have their own formulas, and that some suppliers are now using breakaway plastic in place of "candy glass."

4. http://www.homepages.ucl.ac.uk/~ucfbanf/general/crystal.htm

5. G. Sgualdino, E. Scandale, D. Aquilano, G. Vaccari, & G. Mantovani, "The Growth of Large Sucrose Crystals from Solution," *Materials Science Forum* 203 (February 1996): 43–46.

6. *International Sugar Journal* 53 (August 1951): 210. http://books.google.com/books?id=kEoXAQAAIAAJ&pg=PA210&dq=%22large+sugar+crystals%22&hl=en&sa=X&ei=phMzU9TjBcbsrAf25IH4Aw&ved=0CEQQ6AEwAA#v=onepage&q=%22large%20sugar%20crystals%22&f=false

7. Iso-Maltitol shares a chemical formula with Maltitol. They both have a cyclic structure of five carbons—all but one with a hydroxide—and an oxygen atom. One of the carbons in the ring is joined to an oxygen, the other side of which is joined to a six-link linear carbon chain. In Maltitol this oxygen is linked to the third carbon in the chain; in Iso-Maltitol it is joined to the first.

8. http://www.globalsugarart.com/clear-edible-sugar-diamonds-by-gsa-p-24120.html

9. http://cakecentral.com/g/i/1869383/fondant-covered-fondant-details-isomalt-lens-tfl/; http://thegreatbritishbakeoff.co.uk/bake-offs/nina-mcarthur-dslr-camera-cake/

10. Cynthia Graber, "Make Edible Fiber Optics," *Scientific American Explorations* (Spring 2001), cited in US Patent 6,416,800 July 9, 2002. The article has been cited and its recipe repeated on many websites, such as http://science-edu.larc.nasa.gov/EDDOCS/RadiationBudget/fiber_optics.html and http://islamic-world.net/children/fiber_optics.htm. This activity was evidently also performed as a "hands-on" science demonstration at Disney World in Florida.

11. Jeff Hecht, *City of Light: The Story of Fiber Optics* (New York: Oxford University Press, 1999), 125–126.

12. http://www.odditycentral.com/foods/this-japanese-water-cake-looks-and-tastes-unlike-any-sweet-youve-tried-before.html; http://en.rocketnews24.com/2014/06/04/this-amazing-water-cake-just-may-be-the-most-delicate-sweet-ever-created/; https://www.yahoo.com/food/weird-cake-looks-like-giant-drop-of-water-88509002826.html; http://www.thatsnerdalicious.com/nerd-cakes/this-cake-looks-like-a-gigantic-drop-of-water/; http://designtaxi.com/news/366206/In-Japan-A-Rice-Cake-That-Looks-Like-A-Water-Droplet-Melts-In-Your-Mouth/interstitial.html/?advertiser=External&return_url=http%3A%2F%2Fdesigntaxi.com%2Fnews%2F366206%2FIn-Japan-A-Rice-Cake-That-Looks-Like-A-Water-Droplet-Melts-In-Your-Mouth%2F http://cooks.ndtv.com/article/show/this-clear-drop-of-water-is-actually-a-dessert-538540.

13. At the 23rd ACM Symposium on Virtual Reality Software and Technology in Gothenberg, Sweden, November 8–10, 2017 Paper [T-23B], "Proposal of Edible Augmented Reality Marker Made from Edible Retroreflector and Its Prototype," https://confit.atlas.jp/guide/event/vrsj2017/subject/2ATE-23/detail?lang=en. See also T. Uji, Y. Zhang, & H. Oku, "Edible Retroreflector," in *Proceedings of the 23rd ACM Symposium on Virtual Reality Software and Technology*, 5 (New York: Association for Computing Machinery, November 2017).

14. YouTube videos of the edible retroreflector in action are available online at https://www.youtube.com/watch?v=W2cDD90yIqs and https://www.youtube.com/watch?v=_82qoPAt5xc. A video showing its production is here: https://www.youtube.com/watch?v=fixbdeeD8tc

15. M. Nomura & H. Oku, "Edible Lens Made of Agar," in *2019 IEEE Conference on Virtual Reality and 3D User Interfaces (VR)*, 1104–1105 (New York: IEEE Press, March 2019).

16. Miko Sato, Yuki Funato, and Hiromasa Oku, "Edible Retroreflector Made of Candy," in *2019 IEEE Conference on Virtual Reality and 3D User Interfaces (VR)*, 1146–1147 (New York: IEEE Press, 2019).

17. There have been numerous articles about edible optics over the years. Here are some not mentioned earlier: Michael E. Knotts, "Optics Fun with Gelatin," *Optics & Photonics News* 7, no. 4 (1996): 50–51; Patrick Bunton, "Edible Optics: Using Gelatin to Demonstrate Properties of Light," *The Physics Teacher* 35, no. 7 (1997): 421–422; Mario Branca and Isabella Soletta, "Construction of Optical Elements with Gelatin," *The Physics Teacher* 41, no. 4 (2003): 249; Mohammad A. F. Basha, "Optical Properties and Colorimetry of Gelatine Gels Prepared in Different Saline Solutions," *Journal of Advanced Research* 16 (2019): 55–65; Oku Hiromasa, M. Nomura, Kumi Shibahara, and Akihiro Obara, "Edible Projection Mapping," in *SIGGRAPH Asia 2018 Emerging Technologies*, 2 (ACM, 2018); Oku, Hiromasa, Takahiro Uji, Yiting Zhang, and Kumi Shibahara, "Edible Fiducial Marker Made of Edible Retroreflector," *Computers & Graphics* 77 (2018): 156–165; Fiorenzo Omenetto and David L. Kaplan, "Edible Holographic Silk Products," U.S. Patent Application 12/999,087, filed June 9, 2011; Maria Helena Gonçalves, Luis E. E. de Araujo, and Varlei Rodrigues, "Lentes de gelatin," *Revista Brasileira de Ensino de Física* 42 (2020); Paul J. Weber and Brian D. Andreson, "Fiber Optic Candy," US Patent #6,416,800, granted July 9, 2002.

25
Thoughtographs and the Stanhope Lens

Thoughtographs burst unheralded into the mainstream public consciousness with the publication of the September 22, 1967, issue of *Life* magazine. The magazine's circulation was over a million copies a month, and those readers were greeted with an article entitled "The Baffling Case of Ted Serios and His 'Thoughtographs': A Man Who Thinks Pictures" by staff writer Paul Welch in the *Science* section. The four-page article featured six photographs—not unusual for what was, after all, a photojournalism magazine. The first image featured a wrought Ted Serios grimacing into a camera, and another showed him, exhausted, in a chair. But the other four photographs were eerie anomalies—a building, a statue, a car, and what appeared to be a Revolutionary War–era soldier with crossed straps. What made them eerie was that the camera was aimed at Mr. Serios when these were snapped. The photos should have contained his anxious expression, but instead they bore out-of-focus images of other objects, tilted askew. The car photo even seemed to have Mr. Serios's eye inserted into the image (Figures 25.1, 25.2).[1]

According to the article, Serios was a 47-year-old former bellhop who had been producing such thoughtographs since 1955. For the article, Serios had met with *Life* photographers who brought their own Polaroid instant camera and their own film, making trickery, if not impossible, at least less likely. The occasion was the publication of Dr. Jule Eisenbud's book *The World of Ted Serios*,[2] in which the Denver psychiatrist and paranormal researcher detailed his work with Serios. The article gives the impression that how the images were created was a complete mystery. The curious Mr. Serios was a "haunted man" who drank to excess and missed appointments, but he had this one peculiar talent.

A very different story emerged at very nearly the same time in another magazine, *Popular Photography*, in its October 1967 issue.[3] "An Amazing Weekend with Ted Serios" was written by David B. Eisendrath, Jr. and Charles Reynolds. Both were professional photographers as well as skilled amateur magicians. The story that emerged from their article was very different. To begin with, they describe how Ted Serios used a tube that he placed between himself and the camera, which he called a "gizmo." In early accounts, the gizmo is a plastic tube, cut from the container provided with the Polaroid film, and containing the squeegee of "fixer" used to finish and protect the developed image. In later accounts, the gizmo is made from the black paper around the film, rolled into a tube and taped together. Not only does the gizmo not appear in the *Life* article, it's not even mentioned. Yet its presence immediately rouses suspicion (as Eisenbud states in his book).[4] Furthermore, Eisendrath and Reynolds claimed to detect characteristic magician's moves as Serios worked with the gizmo, making them suspect that he was putting something into it to produce his images, and then smoothly and virtually undetected, removing it when presenting the gizmo for inspection.

Figure 25.1 "Thoughtograph" of a car, supposedly created by the power of Ted Serios's mind acting on a Polaroid camera.
Courtesy of University of Maryland, Baltimore County.

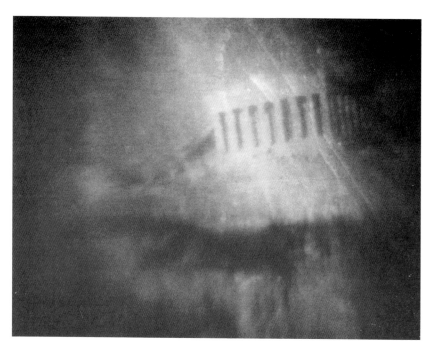

Figure 25.2 "Thoughtograph" of the Parthenon, supposedly created by the power of Ted Serios's mind acting on a Polaroid camera.
Courtesy of University of Maryland, Baltimore County.

They show how a device which can fit into the gizmo that will produce images can be made. All that is needed is a tube with a lens at one end and a photographic transparency at the other, so that the transparency is placed at the focus of the lens. If such a device is illuminated from behind (by, for instance, light reflected from a face—the Polaroid cameras Serios preferred use flash attachments) and the camera is focused at infinity, one will photograph what is on the transparency. Of course, if the focus is not set to infinity, or if the transparency is not quite the correct distance back, the image will be blurred. And it will be likely that the image on the transparency will be rotated relative to the camera's axes (Figures 25.3, 25.4). (One can also place a mirror or retroreflector behind the transparency, so that light from the front will illuminate the image.)

This is precisely what happens in the case of Serios's photographs. If one is not committed to the notion of paranormally created photographs, this is a very likely way to produce them.[5] Some photos from a shoot show Serios with his gizmo in direct contact with the camera lens, and Eisenbud's book describes it being used that way.

Eisenbud was dismissive of such devices and downplayed the capabilities of sleight of hand. He describes them himself in his book (which must have been written before Polaroid's experts, or Reynolds and Eisendrath did their work). He describes a device using a transparency and a "lens of high refractive power interposed" and says it would have to be placed very close to the camera lens. If it were not, then "other and easily identifiable parts of the surrounding scene would be imaged onto the film in addition to whatever was produced by means of it."[4] Significantly, there have been cases where Serios tried to project his thoughtographs onto a video camera, where he was clearly not in contact with it.[5] They show precisely what Eisenbud describes.

It is clear that, if such a device or devices were used (allowing Serios to produce more than one image per session), it implies that he was an intentional illusionist, which further implies that we ought to take nothing he says or implies at face value. His pose as a hopeless and continual drunkard with low education and few skills is itself immediately suspect as subtle misdirection. Just because Eisenbud believed it and

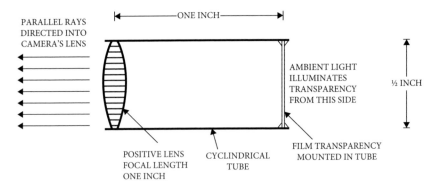

Figure 25.3 Probable construction of the device used by Serios and concealed in his "gizmo" to create his "thoughtographs."

Illustration from *Flim-Flam!* by James Randi.

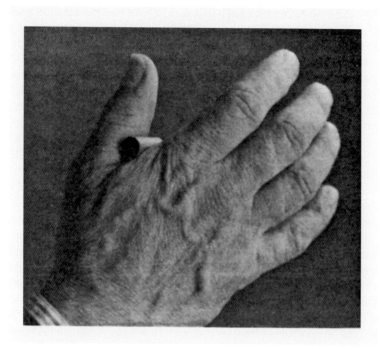

Figure 25.4 Probable method of concealment of the device used by Serios and concealed in his "gizmo" to create his "thoughtographs."
Illustration from *Flim-Flam!* by James Randi.

reported it does not make it true. Magicians have committed themselves to a lifetime of false circumstances in order to effectively work illusions before.

Serios had committed believers to this day, and I apologize to them, but I'm going to accept as probable that his illusion was produced in precisely this way. My whole point in bringing this up, in what is supposed to be an optics column, is to raise a question that I have not seen posed before. Optical engineers know that you can place a reticle, or a pinhole, or even a transparency at the focal point of a lens, and it will act as if that object is effectively infinitely distant. If you view it or image it with an optical system set to look at infinitely distant objects, you will get a sharp image at the focus of your imaging lens. This is how collimators work, for testing optical systems. We use them to test cameras and targeting scopes, and for performing Modulation Transfer Function measurements. But this is hardly common knowledge to people who don't regularly use optics. How did Serios know enough to enable him to construct his devices (which, in deference to his own nomenclature, I will henceforth call "gizmos")?

My first thought was, if we take the working hypothesis that Serios was no stranger to the methods of magicians (as evidenced by his observed sleight of hand), then he might have found the trick listed among the many illusions described or catalogued in the shadowy network of magicians' publications. There are a great many books with limited press runs, and magazines devoted to magic, but these have a limited market and rarely show up in mainstream bookstores. I have made a search of likely texts devoted to optical magic and have drawn a blank. My search was by no means exhaustive, and I can't say that such an illusion does not appear in any written account, but I have not yet located one.

The next place to look was in popular accounts of optics, and here I had somewhat better luck. A number of pieces about the use of optics have been written by Samuel Irvine "Sam" Brown (1906–1976). He wrote for the magazines *Popular Mechanics* and *Popular Science*, edited "The Deltagram" for Delta Manufacturing, and wrote and contributed to books on optics distributed by Edmund Scientific (now Edmund Optics, but which at the time specialized in surplus optics, much of it sold to home experimenters).[6] Significantly, Brown was also an amateur magician and wrote articles describing how to work tricks. He wrote a book on how to make optical illusion devices using the sort of surplus items sold by Edmunds.[7]

Nothing I have yet found among his writings indicates that he describes or suggests a trick of the sort Serios used, but his books do describe the construction and use of collimators at length, with one book—*Collimating Systems* from 1955—being devoted completely to collimators. It shows how a reticle placed at the focus of the collimating lens will be perfectly imaged by a telescope focused at infinity. The diagram of the collimator, minus the light source and diffuser, looks exactly like the diagrams of a "gizmo" from magicians, suggesting explanations for Serios's illusion. *What to Make with Edmund Chipped Lenses* from 1966 shows a Slide Strip Viewer on p. 11 that places photographic transparencies at the focal point of a lens for easy viewing. On the facing page is a diagram for the use of "Supplementary Lenses for Close-up Photography."

Of course, Serios could not have been using information from a 1966 booklet to work his illusion earlier (he had been doing it since the early 1960s, when the paranormal magazine *Fate* started writing about him. His own claim is that he had been doing this since 1955), but the Slide Strip viewer is suggestive.

An optical novelty was developed in 1857 by French photographer and inventor René Dagron. Six years earlier John Benjamin Dancer had invented microphotography, producing photographs as small as 3 mm². Such a photograph required a good microscope for viewing, and these were expensive. Dagron modified a miniature Stanhope lens, which consisted of a very small biconvex lens that acted as a powerful magnifier. He made a plano-convex lens with one focus at the plane surface and then affixed the microphotograph to the flat end with Canada balsam. A person holding the arrangement up to his or her eye and looking in the curved end, with the transparency toward a light source, could see the microphotograph greatly enlarged.

He called these "bijoux photo-microscopiques" or "microscopic photo-jewelry" and displayed them at the 1859 Exposition in Paris and at the 1862 International Exposition in London, where it received an honorable mention. The miniature photographs with built-in magnifiers became immensely popular. They could be mounted in all sorts of unlikely and unexpected places, such as the handles of canes, where the viewing hole might not be obvious, and could be made to hold images of loved ones or risqué images. Almost by design, the images were secret and limited to one viewer, which suggested both possibilities. They were sold with photographs of famous attractions, as souvenirs that could be savored long after the visit.

In the 1950s, cheaper versions of such Stanhope viewers started to be made for the postwar tourist trade, using inexpensive tiny photographs and cheap injection-molded plastic lenses. Because these were often of poor optical quality, they were thin and not placed in contact with the lens. They were, however, still placed one focal length from the transparency.

It would not be difficult to imagine someone playing with one such viewer and, noticing how it presented an apparently enlarged image in a small, easily concealed source, then taking the logical step of placing it against a camera lens and snapping off a picture. Once that was done, a little experimentation would reveal how to construct one from scratch, producing a larger photographic image by using a larger diameter lens. I am delighted to find that the webpage of Nile Root, who attended Serios's performances in the 1960s, agrees not only with Reynolds, Eisendrath, and Randi but also concurs with the inspiration of Serios from "Stanhope"-type souvenir viewers of the 1950s.[8,9]

Notes

1. https://books.google.com/books?id=QlYEAAAAMBAJ&pg=PA114&dq=%22ted+serios %22&hl=en&sa=X&ved=0ahUKEwiVx_yMzfPLAhVM2RoKHTsaAlwQ6AEIKDAD#v= onepage&q=%22ted%20serios%22&f=false
2. Jule Eisenbud, *The World of Ted Serios: "Thoughtographic" Studies of an Extraordinary Mind* (William Morrow and Co., 1967). Besides Eisenbud's book, he writes about Serios and thoughtograpohy in *Psychic Exploration: A Challenge for Science*, Edgar D. Mitchell and John White, eds. (Perigee Books, 1974), chapter 13, "Psychic Photography and Thoughtography," 314–332.
3. David B. Eisendrath, Jr. and Charles Reynolds, "An Amazing Weekend with Ted Serios," *Popular Photography* (October 1967): 81–87, 131–141, 158. Eisendrath and Reynolds later appeared on Arthur C. Clarke's *World of Strange Powers* on The Discovery Channel to talk about their experiences with Serios: https://www.youtube.com/watch?v=uETwx6HwfQU https://www.youtube.com/watch?v=Pm3DX96ILjw. Magician and skeptic James "The Amazing" Randi has written about Serios numerous times, at greatest length in his book *Flim-Flam!* (New York: Prometheus Books, 1982), 222–228, where he gives a diagram of a "gizmo," shows one way to conceal it, and shows a photograph made using such a device. Skeptic Martin Gardner discussed the case in "A Skeptic's View of Parapsychology," *The Humanist* (Nov/Dec 1977). Reprinted, with many additions, on pp. 141–150 of *Science: Good, Bad, and Bogus* (New York: Avon Books, 1981).
4. Jule Eisenbud, *The World of Ted Serios: "Thoughtographic" Studies of an Extraordinary Mind*, 107.
5. The *Life* magazine article even hints at this. Referring to a team of experts from Polaroid called in to look at the results, it says "one theory was that that he somehow used miniature negatives to produce his effect." Stated baldly like that, it seems an unlikely grasping at straws. But if the Polaroid team had suggested a device such as Eisendrath and Reynolds describe—and I suspect that they did—it seems eminently plausible. There exists film of Serios "creating" an image, which is incorporated into an episode of the television show *In Search Of . . .* ("Ghosts in Photography," originally broadcast October 5, 1981). Watching this film, you get glimpses of an out-of-place image that seems to originate from Serios's

hand. To the average viewer, it seems inexplicable, but it so clearly appears to be the result of the collimator device described in the text that it eliminates any doubts in my mind.

6. One is currently posted on YouTube at https://www.youtube.com/watch?v=5dJrI1X2Z4U

7. Some of Brown's books are available for free download through Anchor Optics, a division of Edmunds: https://www.youtube.com/watch?v=5dJrI1X2Z4U

8. See   http://www.circuitousroot.com/artifice/resources/biblio/sam-brown/index.html   for probably the fullest biography and bibliography of Brown

9. http://www.niler.com/estitle.html. His diagram of the device inside Serios's gizmo resembles Randi's.

26

The Best Disinfectant

Sunlight is said to be the Best of Disinfectants.
—Louis Brandeis, "What Publicity Can Do," in *Other People's Money—
and How Bankers Use It* (1914)

When Louis Brandeis wrote those words, two years before he was appointed as a justice of the United States Supreme Court, he was speaking metaphorically. But the statement derives its force from the commonplace knowledge and belief that sunlight *was*, in fact, a readily available, free, and effective disinfectant. Certainly there is a tradition of cleansing through exposure to sunlight that runs back to the ancient world. This collective knowledge was almost certainly derived from observation, but a real understanding of the effect had to await the discovery and acceptance of the germ theory of disease. Immediately after the work of Louis Pasteur came two papers by the British scientists Arthur Downes and Thomas P. Blunt.[1] In these experiments, the two scientists used Pasteur's solution as a growth medium for bacteria. They exposed the specimens to sunlight and observed the destruction of the bacteria. They concluded that "Light is inimical to the development of *bacteria* and the microscopic fungi associated with putrefaction and decay." In the second paper they continued this work, crudely examining the dependence upon wavelength by using colored glass filters to screen the light. They found that blue and violet light was more effective than others, but that red was not ineffective.

Although others continued the work, the next successful step didn't come until 1892–1893, when H. Marshall Ward, professor of botany at the Royal Indian Engineering College at Cooper's Hill in Surrey, published three papers extending these experiments.[2] He began by reproducing the work of Downes and Blunt, employing the more modern agar instead of solutions as a growth medium. He used a prism to throw a continuous spectrum of sunlight across the agar, thus giving him a direct image of the effect of different wavelengths of light on the growth of the bacteria. The bacteria, in fact, acted like a photographic medium, and he could see the dark Fraunhofer lines of the solar spectrum standing out as regions of live bacteria against the dead areas that had been irradiated. This gave him a calibration for his wavelengths, and he could see that growth was inhibited in the blue and violet regions.

He decided to push on beyond the ultraviolet (UV) wavelengths that sunlight could give him by using an electric lantern and a quartz prism, and he observed that the effect extended well into the ultraviolet, confirming for the first time that these deeper ultraviolet rays were much more effective at killing the bacteria than visible light.

In 1917, H. S. Newcomer, a doctor at the Henry Phipps Institute at the University of Pennsylvania, made a series of experiments using an iron arc light where he tried to better characterize the useful region of germicidal rays.[3] Remarkably, given the limitations of his equipment, he was able to find that the efficiency peaked at about 260 nm. After much more experimental work over the succeeding decades, it was found that the peak germicidal value for ultraviolet light is at 265 nm, and that it is due to ultraviolet light breaking bonds in DNA molecules. These recombine improperly, a process called thymine dimerization. The resulting DNA has too many flaws to operate effectively, and the organism dies. That peak is close enough to the strong 254 nm line of a mercury lamp (with a quartz envelope to transmit that light) that use of a mercury lamp for germicidal applications became the standard. Xenon arc lamps and flashlamps and ultraviolet LEDs can now also provide that radiation.

I have to admit that, once I learned that deep ultraviolet light was responsible for killing bacteria, fungi, and even virus through DNA disruption, it seemed as I understood the purifying action of sunlight. But a little thought shows that this cannot be correct—the earlier work used sunlight filtered through the atmosphere, where the ozone absorbed the deep ultraviolet light. The work of Downes, Blunt, Ward, and other researchers of the period demonstrated quite clearly that, although its germicidal power was not as great as that of the deep ultraviolet available from artificial sources, ambient sunlight throughout the near ultraviolet and visible was still an effective germicide. It's true that the 265 nm germicidal peak has "tails" that extend off to longer wavelengths, and the germicidal action persists into the red and beyond. Clearly something else is at work.

The scientific and water supply communities did not abandon the concept of visible light purification of water after it was found that ultraviolet was more effective. Calling the process "insolation," they quantified it and continued to use it in the treatment of water. It was recognized, however, that insolation was a much slower process than UV or chemical sterilization, and communities came to rely upon the much more rapid and thorough treatments to cleanse their drinking water.

What changed the status of insolation was a set of circumstances in which the standard means of purification were not easily available. Dr. Afrit Acra of the American University of Beirut was working with the World Health Organization (WHO) in the late 1970s to expand treatment of diarrheal disease, a major health problem in the developing world. They were promoting the use of oral rehydration solutions (ORSs) that would replenish the body's supplies of glucose or sucrose, along with sodium and potassium salts that become depleted by diarrhea, especially in infants. The problem was finding clean water to use in creating this solution in areas not supplied by clean water, or even by electric power. They discovered (some say by accident) that by sealing water in a polyethylene terephthalate (PET) water bottle, of the sort usually used to sell water and soft drinks, and exposing it to sunlight for a period of six hours or so, the action of the sunlight would destroy the bacteria in the bottle, including the pathogens.

This outcome seems counterintuitive—the Earth's atmosphere itself blocks the deep ultraviolet, and the walls of the plastic PET bottle will further block these. Placing the bottle in the sun will raise its temperature, but not enough to kill bacteria. In fact, one would think that in the warm bottle, with any remaining nutrients trapped inside, the bacteria would find a congenial environment to prosper, reproduce, and grow, just like those in the test tubes of Downes and Blunt. Instead, the opposite occurs. Acra and his associates performed many tests to verify this and published their results in the influential medical journal *The Lancet* in 1980.[4] Four years later, after much more testing, the World Health Organization published a pamphlet about the use of this procedure for purifying water.

The procedure has since acquired the name SODIS, for *solar disinfection*. It is still widely recommended and still faces the uphill battle of disbelief. It seems so unlikely that so simple a process using low technology and easily available materials can be so effective. But SODIS literature shows people scooping up water from even the most unlikely of sources and laying the filled bottles out in rows upon metal or highly reflective surfaces (which let the light pass through multiple times). The main requirements are that the water be as clear as possible (to facilitate passage of the rays) and free of chemical contamination (the process will kill bacteria, but it won't remove poisonous chemicals).

How does it work? Much work has been done on examining the reactions caused by sunlight on water to create activated oxygen compounds. These, it is felt, are what are responsible for killing bacteria in natural waters and were behind the sterilization observed in the late 19th and early 20th century. The process is still imperfectly understood. It should also be pointed out that sunlight has been shown to kill bacteria and fungus under completely dry circumstances.[5] Evidently there is also some direct effect of the light upon the bacteria, without the moderating effect of water.

There have been variations on the SODIS formula, using special glass reactors with focusing mirrors to direct more light at the water, or devices that use ultraviolet LEDs to deliver the UVA light used to process the water. But these seem fundamentally at odds with the circumstances that make SODIS the most useful—its very low-tech nature means that it can be used anywhere you can get PET soda bottles, without specialized and often bulky equipment, and without electricity from a generator, or even a battery. There are no moving parts to break, no chemicals to replenish, no batteries to replace. All you need are old plastic bottles with screw caps.

Today several groups and organizations are making available literature on the use of SODIS to provide clean water to parts of the world that still lack it, including the Centers for Disease Control and Prevention in the United States, the World Health Organization, UNICEF, Red Cross and Red Crescent, and the Swiss Federal Institute (EAWAG).

Notes

1. Arthur Downes and Thomas P. Blunt, "Researches on the Effect of Light upon Bacteria and Other Organisms," *Proceedings of the Royal Society of London* 26 (1877): 488–500; and "On the Influence of Light Upon Protoplasm," *Proceedings of the Royal Society of London* 28 (1878): 199–212.
2. H. Marshall Ward, "Experiments on the Action of Light on Bacillus anthracis," *Proceedings of the Royal Society of London* 52, (1892): 393–400. H. Marshall Ward, "Further Experiments on the Action of Light on Bacillus anthracis," *Proceedings of the Royal Society of London* 53, (1893): 23–44. H. Marshall Ward, "The Action of Light on Bacteria. III," *Philosophical Transactions of the Royal Society of London. B* 185 (1894): 961–986.

3. H. S. Newcomer, "The Abiotic Action of Ultra-Violet Light," *The Journal of Experimental Medicine* 26, no. 6 (1917): 841–848.
4. A. Acra, Y. Karahagopian, Z. Raffoul, and R. Dajani, "Disinfection of Oral Rehydration Solutions by Sunlight," *The Lancet* 316, no. 8206 (1980): 1257–1258.
5. See, for example, C. Phillip Miller and Doretta Schad, "Germicidal Action of Daylight on Meningococci in the Dried State," *Journal of Bacteriology* 47, no. 1 (1944): 79.

27
Walter Darcy Ryan and His
Electric Scintillator

On the evening of Thursday, July 12, 1906, something strange was going on at the
Relay Yard near Bass Point in Nahant, Massachusetts. Crowds gathered in anticipa-
tion, directed there by an advertisement in the Lynn *Daily Evening Item*. A group of
men from General Electric Company in nearby Lynn, across the Causeway, had gathered to set up a peculiar device in the space between the Relay Hotel and the loop
ending the railway. It consisted of a tall Christmas tree–like structure of pipes carrying
steam from a generator and a bank of five searchlights on a platform, each manned by
an operator who could swivel it in any direction, and a set of colored gelatin filters that
could be placed in front of them.[1]

After dark, the generator was fired up and steam sprayed out from nozzles at the
ends of the pipes, rising to a height of 40 feet. At 9:30 the colored searchlights were directed at the clouds of steam, illuminating them to create streaks of brilliantly colored
light. The cloud could be all one color or streaked with several colors. By moving the
searchlights in tandem, they could create striking patterns, including ones that suggested the flags of different nations. Crossing differently colored beams mixed them,
creating new colors. By using special rotating nozzles, they could create pinwheels of
light, or "snakes" and "ghosts." They could interrupt the streams of steam with thrown
objects, add smoke to it, and set off fireworks within the steam. There were many different effects and patterns. The operators of the searchlights were signaled using whis-
tles and a hand flashlight.

The *scintillator*, as it was called, was the creation of Walter D'Arcy Ryan. Ryan was
born on April 17, 1870, in Kentville, Nova Scotia. He trained for a career in the military but found himself drawn to engineering and electrical devices. He emigrated
to Lynn, Massachusetts, and joined the Thomas-Houston Electric Company, which
that year merged with Thomas Edison's company to form General Electric (GE). On
December 18, 1903, he became GE's first illumination engineer. In his years of service
he designed street lamps, automobile headlamps, and other illumination devices, performing calculations and measurements to produce the ideal forms.

But he was also interested in aesthetically appealing architectural lighting. The
recent World's Fair and Exhibitions, such as the 1893 World Columbian Exposition
in Chicago, were showcases for the new capabilities of electric lighting. GE had lost
the contract to light the 1893 exhibition to Westinghouse, and Ryan set out to create
something spectacular and different for the upcoming Jamestown Tercentenary
Exhibition to be held in Virginia. The device set up at Bass Point in Nahant was to be a
demonstration of the dramatic effects that GE was capable of.

They had initially hoped to continue the light show for two weeks, if they could.
They ended up performing it on that site for 54 consecutive days. People chartered

boats from Boston to come out to Bass Point to watch. The proprietors of the newly opened Wonderland Amusement Park in Revere, on the other side of Broad Sound four miles away, must have fumed with frustration at this upstart attraction that stole interest and customers away from their fledgling venture. At first they tried drowning out the effect by shining their own powerful searchlight across the sound at it. When that didn't work, Floyd C. Thompson, the general manager of Wonderland, did the only reasonable thing—he bought the use of the scintillator for the remainder of the summer, shipping it across to Revere Beach. "This is the crowning glory of Wonderland," announced the coverage in the papers.

That the scintillator was an immediate hit was proven by its appearance almost immediately in a book, *Looking Forward* by H. W. Hillman, intended as a sequel to Edward Bellamy's 1888 future utopia novel, *Looking Backward*.[2] The book purports to be written in the then-far future of 1912, telling of all the marvels achieved by electricity since Bellamy's time and for the six years from its actual writing. It waxes eloquent about the scintillator:

"There comes the scintillator," said Ethyl. Just then the lights were turned low on the pavilion, and a series of five searchlights sent their streams of illumination into view. Steam pipes were located some thirty feet in the air, and as the many jets emitted forth their volumes of steam, the various colors from the scintillator made beautiful clouds of all shades, ever changing as the signals for proper color combinations were given. The policy of the proprietor of this Mountain House, was to furnish night illumination scenes from the scintillator at least one night each week, and the display apparatus was always on exhibit in connection with holiday festivals.

"I remember," said Tom, "when this scintillator was invented, and commercially introduced. It was only a few years ago. One of the first exhibitions was given at a seashore resort near Boston, and it was exceedingly popular. Enormous crowds came to the beach on the trolley cars and automobiles, to spend the evening, and witness the beautiful illumination displays. The newspapers commented very favorably upon the invention, and prophesied even at that time that scintillators would be adopted by all up-to-date summer resorts, mountain houses, and wherever the people congregated for an outing, or an evening of pleasure. I remember distinctly, attending the Jamestown Exposition, where 100 searchlights were used in connection with a grand illumination scheme. It was the most marvelous exhibit of illumination which I had ever witnessed." Rhine, the inventor of the scintillator, and his wonderful reputation as an illuminating engineer. He said that not more than a year after the first exhibition of the scintillator near Boston, Prof. Rhine's engineering ingenuity had created various schemes representative of the introduction of fireworks displays, and that the next 4th of July he produced the most spectacular exhibit of fireworks by means of the electrical scintillator, without using powder, or any of the ordinary devices which had been common for years in connection with fireworks exhibitions. The results of this wonderful display led electrical papers, and the public press throughout the entire country, to disseminate information about the wonderful illumination scheme, so that the following year, electrical companies in general arranged their plans so that electricity became the agent for 4th of July demonstrations, and the old style fireworks were abandoned in connection with large demonstrations and illumination exhibitions.

Hillman curiously misspells Ryan's last name as "Rhine," as if he had only heard it pronounced, not written down. He was so taken by the scintillator that an image of it in use is used as a frontispiece for the book. Unfortunately, his predictions didn't come to pass. The scintillator was not widely used at resorts and amusement areas, and it did not replace traditional fireworks. It did not even appear at the Jamestown exposition, for which it had been designed and built. The plan had been to put on shows at night, with the scintillator depicting the flags of participating nations. But the Exposition had financial difficulties, and the scintillator was pulled from its show.

The scintillator found a different use that year, however; it was used to illuminate Niagara Falls.[3] (This wasn't the first time the Falls were illuminated—the history of illumination goes back to 1870—nor the first time they were illuminated in color, a practice that continues to this day. Some sources claim that this is, however, the first time the entire Falls were thus illuminated.) For this application 31 spotlights, each 30 inches in diameter were used, arranged in two batteries, one of 5 lamps and one of 25 lamps.

Ryan's scintillator finally got a full public demonstration at an International Exposition at the Fulton-Hudson Celebration in 1909. For this exposition, Ryan set up lights at Riverside Drive and 155th Street, overlooking the Hudson. It was now called the "electric steam scintillator." It used two batteries of searchlights, each with nine 30-inch lights and one 24-inch light. The pipes, supplied with steam from a 200 horsepower boiler, ran up to 100 feet high, and were placed 100 feet in front of the lights. The program lasted an hour and a half, and "included sunbursts, auroras, peacock feathers, beam dances, plumes, etc. In addition to the steam effects, bombs were exploded."

The scintillator's biggest showcase, however, was at the Panama-Pacific International Exposition in San Francisco, California, in 1915. This was the crowning glory for Ryan, because he was the director of illumination for the exposition. He designed the illuminations for the buildings, commissioned a "Tower of Jewels" covered with high-index glass jewels illuminated by sunlight by day and searchlights by night, and he brought out the largest incarnation of his scintillator yet.

He arranged for the use of an entire steam locomotive from Southern Pacific to provide the steam and constructed a double-level platform for the 48 searchlights, each equipped with seven different colored gelatin filters, which were coated with spar varnish and turpentine as protection against moisture, all of it placed on a double-deck pier on the North breakwater of the yacht harbor. Fireworks mortars, a "chromatic wheel," and a fixture for the "fighting serpents" display were nearby. To coordinate the searchlights, a company of marines was used, responding to signals.

This incarnation of the scintillator really caught the attention of people, and it was immortalized in color in postcards, in rare colored articles in magazines and journals, and in the souvenir booklet for the Fair. Previous articles on the scintillator had only used black and white images, and this did not convey the impact. Poet Edwin Markham (his poem "Lincoln" would be read at the dedication of the Lincoln

Memorial), said of the scintillator, "I have tonight seen the greatest revelation of beauty that was ever seen on the earth" (Figure 27.1).

This was the apex of fame for the scintillator. There were two more uses of it, however. When the Union Trust Building was built in Detroit, Ryan supplied colored searchlights for the roof. The installation consisted of eight 36-inch colored searchlights. Their motion, as the name implies, was controlled by motors, rotating horizontally while oscillating vertically. The lighting at Niagara Falls was rebuilt in the 1920s, and Ryan was in charge, but the resulting system wasn't called a "scintillator."

The scintillator came out for one last showing for the 1933 Century of Progress International Exposition in Chicago. Ryan, again, designed the illumination. As before, there were two banks of lights, twelve 24-inch and twelve 36-inch searchlights, located on the shore of Lake Michigan just below the Travel and Transport Building.

It was the last hurrah for the scintillator. Ryan died of a heart attack on March 14, 1934. Without his impetus, there was no further call for the scintillator, which had appeared at every International Exposition held in the United States (except for the ill-fated Jamestown exposition) since it was constructed. It did not appear at the 1939 World's Fair in New York. The scintillator in San Francisco lived on for many years but was moved from place to place. The lights atop the Union Trust Building—now renamed the Guardian Building—were turned off in 1940 during World War II and were never turned back on.[4] (An effort was made to replace the now-missing lights in the restoration of 1988, but it was deemed too wasteful in the aftermath of the oil crisis.)

Other celebrations copied Ryan. For the welcome-home ceremonies following the return of US troops from World War I, an Arch of Jewels, echoing the Tower of

Figure 27.1 Postcard of the Ryan Scintillator from the 1915 Panama-Pacific Exposition held in San Francisco. You can see four of the steam-producing devices to the right.
Image reproduction by S. Hoyt.

Jewels, was set up at Fifth Avenue in New York City, illuminated at night by multiple searchlights, but Ryan was not involved. There had been a static arrangement of vertical searchlights around the Eiffel Tower in the 1930s. But by the late 1930s quite a different association came about for banks of searchlights. Hitler's architect Albert Speer, pressed for time to erect an arena for the 1933 Nuremburg rally, had opted for a quick solution, using 152 uncolored anti-aircraft searchlight set 12 meters apart and pointing upward. He called it the *Lichtdom* ("Light Dome"), although it usually called in English the "Cathedral of Light." Necessity had resulted in this dramatic display, which was featured in the 1937 propaganda film *Festliches Nürnberg* and used for the closing ceremony of the 1936 Berlin Olympic Games. Mussolini copied the effect, as well, and it became irrevocably associated with totalitarian regimes. Orson Welles thus used it in his 1937 Mercury Theater production of *Caesar*, a modern-dress production of Shakespeare's *Julius Caesar*.

The idea of using lighting effects for entertainment gradually worked its way back in the liquid light shows of 1960s clubs and discotheques, followed by the light shows of the 1970s and the laser light shows that followed shortly thereafter. The light shows of modern music presentations and the laser shows use computer-controlled motion involving galvanometric motors and piezoelectric manipulators that replace Ryan's batteries of human operators (for all the "futuristic" appeal of Ryan's scintillator, it's ironic that, except for the ones atop the Guardian Building, they were all worked by hand), but these shows are limited to interior performances and small venues. The time is arguably ripe for a revival of the Ryan scintillator, with its steam batteries, moving piping, smoke generators, and pyrotechnics now supplemented by computer-controlled addressable mechanical motion.

Notes

1. *Lynn (MA) Daily Evening Item*, July 10–17, 1906.
2. Wikipedia lists 27 unofficial sequels to Bellamy's book and a number of rebuttals as well. https://en.wikipedia.org/wiki/Looking_Backward
3. This wasn't the first time the Falls were illuminated—their history of illumination goes back to 1870—nor was it the first time they were illuminated in color, a practice that continues to this day. Some sources claim that this is, however, the first time the entire Falls were thus illuminated.
4. An effort was made to replace the now-missing lights in the restoration of 1988, but it was deemed too wasteful in the aftermath of the oil crisis.

28

The Trilobite's Eye

The race of man shall perish, but the eyes
Of Trilobites eternal be in stone,
And seem to stare about with wild surprise
At changes greater than they yet have known
—Timothy Abbott Conrad (1840)

The eyes of trilobites are extremely interesting. They come in at least three structures—holochroal, schizochroal, and abathochroal—and possibly more, depending upon whom you talk to. Some trilobites were eyeless, like *Trimerocephalus*, while some, like *Neoasaphus*, had eyes on stalks, like snails. Others, like *Erbenochile issomourensis*, had immensely tall, columnar eyes. Trilobites lived over a period of 300 million years, across the Paleozoic era from the Cambrian to the Permian, then disappeared before the dinosaurs arrived.

Although there are fascinating issues about the various designs of trilobite eyes, the forms of their lenses, and other details, in this brief piece we are concerned with only one aspect of the trilobite eye. Its external lens was made of the mineral calcite, a trigonal form of calcium carbonate. One single crystal form of this mineral is Iceland spar, well known for its pronounced birefringence. Trilobites are almost unique in having calcite eye components—some modern brittle stars do, as well as some ostracods.

This is a very odd thing—why would trilobites have mineral components in their eyes? Of course, stating it that way is misleading. Trilobite eyes have calcite lenses in the same way that humans have hydroxyapatite formations for teeth. In both cases, the "mineral" is really a combination of the mineral components and biological material, all produced by the body of the creature. In both cases, the material originally formed the skin or exoskeleton of the animal. Human teeth are very similar in structure to shark skin, and it's clear that some early common ancestor repurposed part of its spiny covering to act as biting and chewing apparatus. In the case of the trilobite, its entire carapace and exoskeleton was largely calcite. An early ancestor of the trilobite clan developed part of that carapace into optical apparatus, making it more optically clear so that it could be used for focusing light onto ommatidia, just as the compound eyes in insects do.

That basic fact—that trilobite eyes have calcite lenses because their carapaces are constructed of what is essentially calcite—is extremely important. It highlights that, although trilobites are arthropods, they are not particularly closely related to modern arthropods (including the horseshoe crab, which is often said to be its closest living relative). Modern arthropods have carapaces made of chitin, a long-chain polymer of *N*-acetylglucosamine, an organic material with no mineral content. This implies that trilobite eyes developed independently of other arthropod eyes—lobsters, crabs, and insects recommissioned parts of their chitinous exoskeletons to form lenses, just as trilobites rearranged their calcite exteriors for the same purpose. No evolutionary process causes a creature to substitute one material in an existing eye for another. Trilobites and other arthropods developed their eyes independently.

Evolution isn't a process by which an optimum design is produced by the processes of adaptive radiation and natural and sexual selection—it's the use of those procedures to create structures that provide some sort of advantage over not having them. That the result isn't necessarily ideal is frequently shown by the ways that organisms, using less-than-ideal starting material, produce what is arguably the best that can be done with that raw material. In the case of the calcite eyes of the trilobite, the biggest problem is perhaps the fact that the crystal, although optically very clear, is birefringent. The very large difference between the ordinary and extraordinary refractive indices leads to problems producing a tight single focus. Remember the way that the calcium carbonate crystal Iceland spar has radically different angles of refraction for two different polarizations. Imagine having to design a single lens out of that material.

The trilobite did. It minimized the separation of images of different polarization by reorienting the crystal so that its "c" axis (along which the two polarizations have essentially the same index) lies along the optical axis of the eye. This minimizes the focal spot. Much has been made of the fact that trilobite lenses are also aspheric, but this is less impressive when you consider that trilobite lenses aren't imaging lenses— they focus light onto the ommatidia and are basically "power in a bucket" systems, although this design will concentrate the light energy onto relatively small detectors.

One other unfortunate property of calcite is that it fluoresces under ultraviolet radiation,[1] and this can make them inappropriate for eyes that have to be used where ultraviolet light is present. It has been argued that trilobite eyes are used in circumstances where most of the ultraviolet light has been filtered out already, so it is of no consequence. But the flip side of this argument is that such fluorescence prevented the eyes from being used where ultraviolet light *was* present. Descendants of other arthropods, with chitin-based lenses, made the transition to living out of the water, in the open air where ultraviolet light is present. No trilobite descendant could do this. It's possibly because trilobites had calcite lenses that they had no terrestrial descendants. As a result, none of the clan of trilobites could survive whatever environmental stress caused the great trilobite extinction at the end of the Permian period.[2]

There are two other forms of calcium carbonate—aragonite and vaterite. The latter isn't terribly stable, but aragonite is. For years it was thought that no animal had aragonite lenses, until one was discovered in a rather spectacular fashion.[3] Daniel Speiser of the University of California at Santa Barbara was studying the lenses of an undersea arthropod with the charming name of West Indian Fuzzy Chiton. He dropped two excised lenses into an acid bath to clean them off, and he found that instead they rapidly dissolved. In retrospect, this is similar to the undergraduate experiment of dropping vinegar or other weak acid onto limestone—both aragonite and calcite are crystalline forms of limestone. Vinegar makes limestone bubble up and dissolve. He hadn't expected the chiton's eyes to be made of the same material as its carapace, which is of the same calcium carbonate that many shells are constructed of. The question of whether a carapace is made of calcite or aragonite is likely tied up to the type of seawater the creature developed in—there are "calcite seas" and "aragonite seas."[4] The chiton eye, by the way, is less than 10 million years old. It represents yet another independent development of the eye, and a fairly recent one at that. So far as I am aware, the chiton's aragonite lens is unique to that creature.

Notes

1. Brigitte Schoenemann, Euan N. K. Clarkson, and Gábor Horváth, "Why Did the UV-A-Induced Photoluminescent Blue-Green Glow in Trilobite Eyes and Exoskeletons Not Cause Problems for Trilobites?" *Peer Journal* 3 (2015): e1492.
2. Euan N. K. Clarkson, "The Evolution of the Eye in Trilobites," *Fossils and Strata* 4, no. 7 (1975): 7–31.
3. Ed Yong, "Chitons See with Eyes Made of Rock," *National Geographic Online*, April 14, 2011. https://www.nationalgeographic.com/science/phenomena/2011/04/14/chitons-see-with-eyes-made-of-rock/
4. Daniel I. Speiser, Douglas J. Eernisse, and Sönke Johnsen, "A Chiton Uses Aragonite Lenses to Form Images," *Current Biology* 21, no. 8 (2011): 665–670.

29
Infinitely Distant

One of the great things about writing a book or an article is that you appear to be incredibly knowledgeable, infinitely well-read, and able to instantly pull answers and obscure references from thin air with great facility. (At any rate, you can seem like that until the letters come in, pointing out your flaws and errors.) This is an illusion created by the fact that all the time spent researching the column and the dead ends and blind alleys gone down are unreported to you, the reader, and are thus invisible. In fact, I am incredibly fallible and frequently wrong, and my saving graces are persistence and an ability to research things in the library and online until I find the answers. I then report the phenomena and their explanation in an interesting and intelligible way (I hope!). Every now and then I am reminded of how dumb I am.

I had such an experience working with an adjustable collimator. I've worked with countless collimators and built a great many of them over the years. You take a tube and place a well-corrected lens at one end. You place a pinhole or a target at the focal point of the lens, and usually a light source to illuminate the target or to send light through the pinhole. Test the separation by looking into the collimator with a telescope set to infinity, or by verifying that the collimated beam stays constant in size.

If you look down into the collimator, through the lens at the target, you see it as if it's located at infinity, but you see a pinhole or target reticule that subtends a finite angle, not blown up to fill the aperture. If you move the target away from the focal position, it goes out of focus when you look at the target through that telescope. Your eyes are literally a bit more accommodating, and you see the target slightly change magnification.

I was trying to explain what was going on to a group of non-optics people. After all, when you have an object at a distance less than the focal length of a lens, the image that you see is erect, a virtual image. That's how a magnifying glass works, after all. When the object moves further away from the lens than the focal length, the image is real and inverted, and it's on the same side of the lens as the observer. Why didn't they see the image of the target flip over once it passed the focal point?

This brought back memories of things I'd been told. You can measure the focal length of a lens by moving it away from, say, a computer screen. As you move the lens away from the screen, the image gets bigger and bigger, until the light fills the aperture of the lens entirely. This, I had once been told, occurs when the lens is a focal length away from the screen. If you continue to pull the lens away, you get an inverted image of the screen.

Another place this behavior manifests itself is with spectrometers. You can tell when the slit of a spectrometer you're aligning is a focal length away from the collimating lens (or mirror), I'd been told, when an illuminated slit makes the entire lens or mirror blaze with light. Both of these situations seem to be incompatible with a collimator that I can look through and see an infinitely distant but not aperture-filling image.

So what's going on?

It finally occurred to me that you had to take the optics of the observer into account as well. Whether that was a human eye or a camera, it had to be part of the analysis. With either an eye or a camera, there's a lens and a detector plane. If you consider the situation as one where an object in the vicinity of the focal point of a lens of focal length f_1 is imaged by an optical system consisting of that lens plus the imaging system lens of focal length f_2 located a distance d_1 away, then you will have an inverted image in the vicinity of the focal point of the imaging system lens, and its magnification will be approximately the ratio of the focal lengths, $-(f_2/f_1)$. If the target is actually at the focal point of the first lens, the image will be at the focal point of the second, and the magnification will be exactly that value. If I move a bit off the focal point in either direction, the image will be inverted, in either case, and it will move by an amount very nearly equal to the square of the magnification times the distance the target moves, and in the same direction. If you're viewing it through a telescope, a relatively small shift in either direction will make the image go out of focus. Your eye can adjust to small shifts, however, so it will stay in focus longer.

So what does happen when you move the target far enough to get an image that blows up and fills the aperture? What is the distance between the target and the first lens, if it isn't the focal length of that lens?

Performing a paraxial ray trace gives the expressions for the final image location and magnification (see Figures 29.1 and 29.2).

The use of the fact that the initial separation $d_0 = f_1 + \Delta_1$ allows us to considerably simplify the algebra, which can quickly grow tedious. If we define $\Delta_2 \equiv d_2 - f_2$, analogous to the definition of Δ_1, then we get the following results for the distance between the image and the focal point of the last lens, and for the magnification of the image:

$$\Delta_2 = -\frac{\Delta_1 f_2^2}{f_1^2 + \Delta_1\left(f_1 + f_2 - d_1\right)}$$

$$M = -\frac{f_1 f_2}{f_1^2 + \Delta_1\left(f_1 + f_2 - d_1\right)}$$

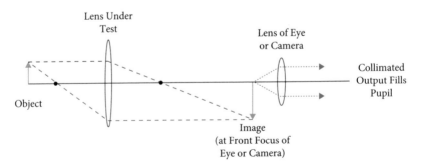

Figure 29.1 Ray trace diagram for object imaged by a lens and viewed by the eye.

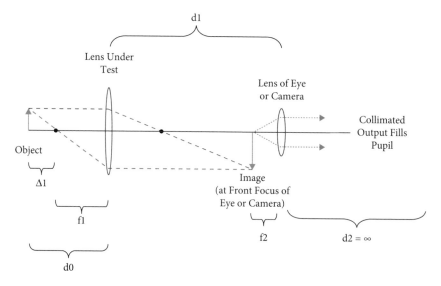

Figure 29.2 Ray trace diagram for object imaged by a lens and viewed by the eye.

Here $d_0 = f_1 + \Delta_1$ is the distance between the target and the first lens and $d_2 \equiv \Delta_2 + f_2$ is the distance between the image and the imaging system lens (either the human eye, with its approximately 20 mm focal length, or a camera). The distance between the two lenses is, again, designated d_1. The image will appear to blow up at infinite magnification, which will happen when the denominator in these expressions goes to zero. This will occur when the distance between the target and the focal point of the first lens is

$$\Delta_1 = \frac{f_1^2}{d_1 - (f_1 + f_2)}$$

It's not immediately obvious what this means, but if $d_1 > (f_1 + f_2)$, then d_0 is greater than f_1, and the target is more than a focal length from its lens. The resulting real image of the target produced by the first lens alone is located a distance $d_1 - f_2$ from the first lens, which is precisely at the focal point of the eye or camera! In other words, we *do* get an aperture-filling infinitely large image with an object at the focal point—only it's when the real image of the target is placed at the focal point of the eye (or camera, if you're using one).

PART III
POP CULTURE

30

I Was a Teenage Optical Engineer

In August 1957, Columbia Studios, through its television licensing division, Screen Gems, made available a package of 52 films, mainly horror film, from their own and other studios to television stations. It was called *Shock!*, and it allowed many of the classic Universal horror films to be shown on television for the first time. Since the films were of high technical quality, visually interesting, very accessible, and relatively inexpensive, they were shown over and over to the TV-imbibing baby boom generation, thereby producing a wave of "monster culture." People of my generation not only knew of Frankenstein, Dracula, and the Mummy, but we associated them with the versions of these made famous by those Universal movies—the flat-topped, bolt-necked Frankenstein monster; the opera-caped Dracula; and so on. They were shared cultural icons that everyone was familiar with. One of the most famous is the Wolf Man or Werewolf, which in its Hollywood incarnation was a very different thing than the traditional *werewolf.*

Upon rewatching the 1941 film *The Wolf Man*, I noticed something very interesting. Lawrence Talbot (played by Lon Chaney, Jr.) was an optical engineer! The son of the Welsh Lord Talbot, he spent several years in the United States, working at an unnamed optical company in California, where he helped install equipment at the Mount Wilson Observatory.

I bring this up because the way optical engineers and scientists are portrayed in popular media—especially motion pictures—ought to be of interest to us in the optics community. There really aren't that many examples of such characters, so it's worth looking at how our profession is portrayed, and what ideas people are taking away from it. In this case, I suspect that most people don't even recall this facet of the character. If anything, they might recall that Talbot grew up in the family mansion. But it certainly is part of the story, with Talbot helping install and align the telescope in the attic-turned-observatory that his father uses for astrophysical research.

In fact, in the original script, the Talbot character isn't even a family member, but an optical technician from Texas named Donald Hill. In subsequent drafts he becomes Larry Gill, but the studio insisted on a familial relationship, so he became Larry Talbot. Like most of the script elements, Donald/Larry and his profession were the creation of scriptwriter Curt Siodmak, a relatively unsung novelist and screenwriter of science fiction and horror who was personally responsible for propagating, if not creating, many of the "traditional" horror tropes, from many of those surrounding the Wolf Man/Werewolf, to vampires dissolving under the influence of sunlight, to the image of a disembodied brain floating in an aquarium. Why, of all things, did he make the man who would become the Wolf Man an optical technician?

Nothing in the script itself directly tells us. Nor does the autobiography of Siodmak, nor earlier versions of the script,[1] nor anything in the lore of werewolves. (Siodmak's script, along with that written by John Colton for the 1935 Universal film *Werewolf of*

London virtually created the entire concept of the Wolfman/Werewolf as people know it today.) Making Hill/Gill/Talbot an optical engineer was a deliberate choice, so why did Siodmak make it?

It's pretty clear from the extant film that he wanted to contrast supernatural forces with modern knowledge. Talbot isn't much of a theorist, as he explains, but he's good with his hands and understands mechanical devices. Later on, frustrated with lycanthropy, he exclaims that he can understand things with wires and tubes (so he's an electronics dabbler, too), but talk of pentagrams and the weird world of the mind are too much for him.[2]

Siodmak's original title of the film was to be *Destiny*, and he wanted to show a man being driven about by the winds of fate.[3] As he originally envisioned it, with Hill not even a family member, it was pure chance that brought him to Wales and his ordeal. Siodmak's original script told a far subtler story than the film as released. His protagonist was bitten by a wolf in trying to defend a woman, and he imagines himself cursed to turn into a wolf. No transformation into a half wolf/half human creature was to be shown on camera, and it was to be left ambiguous whether any supernatural transformation was even taking place. Much of this survives into the script as filmed, with people talking about attacks by a "wolf" rather than a wolf-like creature, and about how the transformation of a human into an animal is impossible, but that it *is* possible for a man to *believe* that he became a wolf. The one moment of Siodmak's original script that showed a monster was to be near the climax, when Hill, pursued by the villagers, pulls himself up to a pool and looks in, seeing himself reflected as a Wolf Man. But, again, was this transformation real, or only in his mind?

Here there might be another reason for his profession: Hill trusts in tools he can understand, and in optics, and it is his optical reflection which reveals to him his bestial nature. But in this important test, can he trust it?

This idea of optics showing one's true self had already appeared in another, largely forgotten short film made by MGM in 1934. It was a tiny Technicolor musical extravaganza entitled *The Spectacle Maker*, based on the 1913 short story "The Magic Glasses" by Frank Harris. In both movie and story, a European craftsman makes spectacles that show people not as they appear to be, but as they really are. Harris's story has a somewhat tragic edge to it, as many parables do, with many of the people viewed appearing grotesque and caricatured by their inner passions. But the short film is bright and cheerful, revealing the innate good in a lame craftsman's son, who goes on to marry a princess. In Hollywood mythology, then, it seems as if optics (and those who work in it) can show us underlying truth.

When *The Wolfman* was remade in 2010, Larry Talbot was no longer an optical engineer, but now a successful actor. That telescope, though, was still there in Talbot House. Only now it was used to look at the moon (which is never even shown in the 1941 film). This film's demons are clearly seen by everyone and don't need optics to reveal them.

Siodmak said in interviews and in his autobiography that he sought out the advice of experts when he wrote about technical subjects. He names them in many cases, but not in this one. I have a suspicion that he talked with someone who worked on the Mount Wilson Observatory, which was not far from Siodmak's Hollywood house. But, unless some other unpublished writings emerge, we'll never know for sure.

My thanks to Edward Coleman of the USC Script Library and to horror film expert Tom Weaver for their help.

Notes

1. Curt Siodmak and Philip Riley, *The WolfMan: The Original 1941 Shooting Script* (Chesterfield, NJ: MagicImage Books, 1993)
2. *Ibid.*
3. Curt Siodmak, *Wolf Man's Maker* (Lanham MD: Scarecrow Press, 2001).

31

Killing Vampires Efficiently
with Ultraviolet Light

That, biscuit boy, is a UV lamp. We're gonna play a game of 20 questions.
Depending on how you answer, you may walk out of here with a tan.
—*Blade* (1998)

That quotation is spoken by Wesley Snipes as the hero Blade in the movie of the same name, threatening a vampire with ultraviolet irradiation if he doesn't answer his questions truthfully. In the past 20 years or so, a new popular trope has emerged, in which ultraviolet light has been identified as the component of sunlight that is responsible for causing vampires to dissolve under solar illumination. Now people are cutting out the solar middleman, so to speak, and using ultraviolet sources as convenient portable devices to kill vampires.

This is a very interesting development. I have written elsewhere about appearances of spectroscopy and knowledge of spectra in pop culture,[1] and I was surprised how few instances there were. But here comes a concept rooted in esoteric knowledge of spectra that grew up in the stew of popular culture entertainment. In essence, the trope that Sunlight Destroys Vampires met up with the trope that Vampires Can Pass Themselves Off as Ordinary People at night, gliding without suspicion through artificially lighted rooms and venues without harm. And what is the difference between incandescent light, candle light, fires, and fluorescent light? It's the fact that natural sunlight has a little more ultraviolet content. Therefore, it's the ultraviolet that is the fatal component. And now that knowledge has been harnessed to create a new weapon against vampires, that has taken its place beside the wooden stakes, acid, crucifixes, holy water, and fortuitous exposure to sunlight.

How did this come about? I don't think it was by someone following the direct logical process. The real story of how ultraviolet light came to be the vampire's nemesis reveals, I think, the way scientific discovery proceeds in the public arena.

To begin with, traditional vampires are *not* harmed by sunlight. We need to dispose of that from the start. This notion didn't exist before the 20th century. What first gave us the idea of vampires dissolving in sunlight was Fritz Murnau's movie *Nosferatu* (1922). The film, an unauthorized adaptation of Stoker's *Dracula*, has a much-condensed but still recognizable plot taken from the novel. But in the film, Count Orlok, the vampire, is closely identified with pestilence and plague. The hero's wife, Ellen Hutter, reads that the vampire can be destroyed if he is detained until sunrise, so she sacrifices herself in order to keep the Count in her room until daybreak. Too late, Orlok realizes that he has been trapped and vanishes in a puff of smoke when sunlight touches him. There may have been a touch of the notion that "Sunlight makes the best disinfectant" in the way the pestilential vampire is destroyed by sunlight.

Many film historians leave it there. Apparently secure in the notion that the concept of sunlight destroying vampires had been introduced in a film, and it then became canon. But vampire films are filled with variations on the standard "myth" that were adopted for the sake of the story or for the beauty of a shot, and which did not become an accepted part of vampire lore.

Florence (Balcombe) Stoker, Bram Stoker's widow, was in deep financial straits when the film was released and was furious that permission had not been asked nor royalties paid. She brought legal action against Prana Film, the makers of *Nosferatu*, and tried to have every copy destroyed. Had *Nosferatu* been the only film using the idea, the trope of the sunlight destruction of vampires would never have caught on.

What changed this, I suspect, was World War II. There was a renaissance of horror films at Universal Studios in the 1940s in which the monsters created in the early 1930s were revived and their stories retold. The fantasy also provided a diversion from the horrors of World War II. And so, when it came to killing the vampires, I suspect that the studios wanted to get away from the bloody impalings that killed Dracula in 1931 and his daughter in 1936. When Universal made *Son of Dracula* in 1943, directed by Robert Siodmak from a story written by his brother Curt,[2] they did something different. The Siodmaks had emigrated from Germany, and Kurt, who changed the spelling to the more American Curt, had probably seen Murnau's film. Dracula's son, played by Lon Chaney, Jr., died by being irradiated with sunlight, which provided a bloodless death for the supernatural creature. Siodmak used the same method to kill off John Carradine as Count Dracula in 1944's *House of Frankenstein*. (Carradine as Dracula died the same way the next year in *House of Dracula*, although Siodmak had nothing to do with that film.)

Siodmak was not the only one to revive sunlight as lethal to vampires. The year 1943 also saw the release of *Return of the Vampire* from the competing Columbia studio. This one starred Bela Lugosi, Universal's original Dracula, as "Armand Tesla," another vampire, but who is, in effect, a revived Dracula. He, too, is destroyed by sunlight at the end.

Again, the idea might have died, but 15 years later yet another studio, Hammer Films, had Count Dracula (played by Christopher Lee) succumb to sunlight at the conclusion of *Horror of Dracula* (1958). Having been used in five films by three different studios, the concept took root, and ever after sunlight dissolution rivaled staking as the preferred method of eliminating vampires, not only in films, but on television, in novels, and in comic books.

How do we get from this to ultraviolet light?

Another outcome of World War II was that a lot of people from northern climates found themselves in places with harsher and longer sunlight, and they were getting sunburned. A lot of research went into developing protective creams to alleviate sunburn. This effort continued after the war. In 1956, a scientist in Hamburg, Germany, Rudolf Schulze, came up with a way to measure and categorize the effectiveness of different protections.[3] He evaluated commercially available protective creams by exposing skin treated with them to light from Osram-Ultra-Vitalux lamps that duplicated the solar spectrum and measuring how long it took to produce noticeable effects. His "Schulze factor" was the first comparative basis for gauging effectiveness.

Another German professor, Franz Greiter of Wittburg, extended the work.[4] An avid mountain climber and skier, he himself got badly sunburned in the mountains and prepared his own protective screens. In 1974, he introduced what was called in English the sun protective factor, or SPF. (Some sources claim that he did so in 1962, but 1974 was the date of his paper.) The SPF was accepted as the basis for rating sunscreens by the United States Food and Drug Administration (FDA) in 1978. The FDA publicized the idea of the SPF and its value.

From ultraviolet rays being harmful to people to ultraviolet rays being harmful to vampires is not a huge leap, with the attendant idea that, if sunblock with a high SPF will protect people, it will also protect vampires. Paul Barber, author of the excellent book *Vampires, Burial, and Death*, which exposed several of the roots of vampiric legends and traditions, in 1996 quoted Stephan Kaplan as saying (at some time before 1996) that "vampires can come out in the daytime; they just need to wear sunblock of 15 or higher."[5] Kaplan, who had founded the Vampire Research Centre in 1972, was speaking mainly about normal people who believed themselves to be vampires, but the phrase could as easily apply to the supernatural beings. The notion was mentioned in *The Magazine of Fantasy and Science Fiction* in 1986.[6] The idea had definitely been planted. Now that the idea of effective sunblock was in the popular imagination, the idea that vampires could profit from it must inevitably follow.

The first fictional use of vampires using sunblock as protection that I am aware of is in the HBO Tales from the Crypt "movie" *Bordello of Blood* in 1996. The idea soon showed up in the *Blade* movie trilogy, and after that, it took off.

As for the idea that ultraviolet light itself is harmful to vampires, the first expression of that idea that I've come across was in the *Rosicucian Digest* (of all places) in 1977.[7] In 1985, it was being casually discussed on *The Science Fiction Radio Show*.[8] It was mentioned in a couple of books in 1990.

By the 1990s, the idea of using an ultraviolet lamp to cause maximum harm to vampires for an investment in electrical power showed up the example from *Blade* cited at the start of this piece. After 2000, the idea started to spread like wildfire. The *Underworld* movie series has a culture of werewolves using "ultraviolet bullets" (which somehow emit ultraviolet light after being shot into a vampire) to destroy their hereditary vampire enemies. Author Christopher Moore has a vampire-fighting kid build ultraviolet LEDs into his jacket in his vampire trilogy, *Blood-Sucking Fiends, Bite Me*, and *You Suck*. The graphic novel *Thirty Days of Night* and its subsequent movie adaptation has people trapped with vampires above the Arctic Circle (where night can last for months—a vampire's ideal location) fighting them off with ultraviolet lamps. The movie *Van Helsing* features a sort of "flash-bang" grenade (probably a magnesium bomb) that eliminates an entire roomful of vampires at once. Ultraviolet light has now been weaponized in the anti-vampire arsenal.

The rise of sunblocks and the effort to educate the public about the teratogenic properties of solar ultraviolet light provided both a new wrinkle in the technological war between vampires and humans (ever since Stoker's *Dracula*, at least, vampire stories have been about supernatural creatures adapting to emerging technology) with the possibility of protective sunblock, and a new explanation for just what it was that was deadly about sunlight—those ultraviolet rays that the high SPF creams blocked.

Notes

1. Stephen R. Wilk, *How the Ray Gun Got Its Zap!* (Oxford University Press, 2013), Chapter 44: "Pop Spectrum."
2. Curt (Kurt) Siodmak was also responsible for much of Hollywood's "werewolf" mythology. See Chapter 30 of this volume.
3. Sergio Schalka and Vitor Manoel Silva dos Reis, "Sun Protection Factor: Meaning and Controversies," *Anais brasileiros de dermatologia* 86, no. 3 (2011): 507–515. http://www.scielo. br/scielo.php?script=sci_arttext&pid=S0365-05962011000300013&lng=en&nrm=iso&t lng=en   See   also   https://www.schrader-institute.de/homepage/testing/testing-skin/ sun-protection/ and Rik Roelands, "History of Human Photobiology," chapter 1 in *Photodermatology*, ed. Henry W. Lim and Herbert Honigsmann, and John L. M. Hawk (Boca Raton, FL: CRC Press, 1999), 1–13.
4. F. Urbach, "Franz Greiter—The Man and His Work" *Photobiology* (Berlin: Springer, 1991), 761. See also *Photodermatology*, p. 768 and F. Greiter, "Sun Protection Factor—Development Methods," *Parf Kosm* 55 (1974): 70–75.
5. Paul W. Barber, "Staking Claims: The Vampires of Folklore and Fiction," *Skeptical Inquirer* 20, no. 2 (March/April 1996): 41–44. https://skepticalinquirer.org/1996/03/staking_claims_ vampires_of_folklore_and_legend/
6. *The Magazine of Fantasy and Science Fiction* (July 1986), 120.
7. *Rosicrucian Digest* 55: 41. https://books.google.com/books?id=jDrZAAAAMAAJ
8. See Daryl Lane, William Vernon, and David Carson, eds., *The Sound of Wonder: Interviews from "The Science Fiction Radio Show* (Phoenix AZ: Oryx Press, 1985), 196.

32

The Endless Corridor

Mirror on Mirror Mirrored is all the show.
—William Butler Yeats, "The Statues" (1940)

One of the most memorable cartoons by Charles Addams (1912–1988) is the one featuring a man getting a haircut in a barbershop that has large mirrors on opposing walls to let the customers see themselves from both sides. The procession of images of the sitter, alternating between forward and backward in a straight and infinite line of figures, is inexplicably interrupted by a figure of a demon in place of the sitter.[1]

The situation in which two mirrors face each other, so that anything placed between them is reflected over and over into infinity, or at least until the images fade due to absorption of the light by the mirrors, is a familiar one. It appears in barbershops and beauty salons, in dance studios and funhouses. It's the physical situation behind "infinity mirrors" and similar illusions. The basic parallel mirror situation lies behind Fabry-Perot interferometers, lasers, White cells, optical delay lines, and multilayer stack mirrors. The artist Yayoi Kusama has been installing mirror-walled (and sometimes floor- and ceiling- walled) "infinity rooms" around the world since 1963. (As I write this in November 2019, one is on exhibition at Boston's Institute of Contemporary Art.)

How old is the effect? Surprisingly, there is no single, unambiguous term for it. "Infinity mirror" dates back only to the late 1970s. In the 19th century a similar construction was used as a popular toy and called the "endless gallery," which consisted of a small box with parallel mirrors at each end. The other walls were decorated like those of a corridor, so the impression one got from looking into a gap at the top of one of the mirrored walls was that of a corridor extending on forever. A related phenomenon is the "Droste effect," where an image contains a tint reproduction of itself, which, in turn, contains a smaller representation of itself. The effect is named after a 1904 advertisement for Droste brand cocoa which exhibited this. It's not quite the same thing, but it does contain the element of recursive elements receding into the distance. "Mirrors within mirrors" captures something of the idea. Sometimes "hall of mirrors" is used, but that term also conjures up mental images of the mirror maze.

This lack of consistent or reliable terminology makes it hard to date the effect. How old is it? Unless someone described it, you might not be able to tell. It might lurk under an unfamiliar terminology. I suspect that the observation of that long tunnel of images formed by making two mirrors face each other dates back to the first time someone had two movable flat mirrors that could be aimed at each other. It doesn't matter if the reflectivity isn't great—two mirrors held close to and parallel to each other will create an illusion quite unlike anything else, with a liveliness and mobility that would have excited anyone's interest. Yet I have found nothing that can be interpreted as a reference to this effect.

Certainly the effect should have been noted by the days of the Roman Empire. Although we associate Roman mirrors with those handheld bronze mirrors, such as have been found at Pompeii, Rome was in fact producing very large and sophisticated mirrors by the start of the Common Era. Pliny mentions mirrors made of glass and backed by metal leaf.[2] Some were made with lead backing. Romans made mirrors large enough to show the entire body. There are references in Seneca and Quintillian.

Some houses had rooms lined with such large mirrors. According to Suetonius, Domitian had such a room, so that he could see what everyone around him was doing (Domitian 14). Suetonius also says that Horace had a room completely lined with mirrors. Claudian describes the Chamber of Venus as covered with mirrors.

Such a mirror-covered room, or even simply a room with large mirrors on opposing walls, must inevitably have given examples of such "infinity mirrors," as in a barbershop.

By the time of the Hall of Mirrors in Versaille (erected 1678–1684), we have sets of large mirrors on parallel walls. Peter Stuyvesant, governor of New York from 1647 to 1664, was supposed to have erected his own little "hall of mirrors," for which he charged admission. Ossian's "hall of mirrors" was erected in Dunkeld, Scotland, in 1783. [3] Nevertheless, no one seems to have commented on the odd optical effects each of these structures must have produced. The earliest comments I've found are in relation to the endless gallery, dating from the late 18th and early 19th century. Giambattista della Porta came close in his 1584 work, *Magia Naturalis*, in which he describes reflection in reflection of mirrors placed at angles to each other, kaleidoscope-style, but doesn't seem to consider the case of parallel mirrors. [4]

There is, perhaps, one earlier reference, but it must be inferred. By about 1590, Elizabethan theater had become enamored of magic. Christopher Marlowe's play *Dr. Faustus*, about the scholar who sells his soul to Mephistopheles in exchange for knowledge, is filled with magic and stage effects. Richard Greene's comedy *Friar Bacon and Friar Bungay* also uses magic effects, including a disembodied, talking head. William Shakespeare had an interest in magic. Not only do several of his plays, such as *A Midsummer Night's Dream* and *The Tempest*, involve magic and magicians, but he seems to have used stage magic in putting on his plays. As magician Teller observed, [5] the stage direction for the magical disappearance of a meal in Act 3, scene 3 of *The Tempest* says that it is caused by "a quaint device," which he interprets as a bit of mechanical stage magic. He was inspired by this discovery to construct his own production of *The Tempest* using all manner of stage magic in 2014.

Magic of a darker sort suffuses Shakespeare's tragedy *Macbeth*, in which three witches prophesy Macbeth's future at two points in the play. In the second of these, [6] they show that Macbeth's reign will not last, and that Banquo's sons will inherit the kingdom. This is illustrated dramatically by the appearance of eight kings, the last of whom holds up a mirror to reveal an infinitely long line of kings.

> Macbeth
> Thou art too like the spirit of Banquo; down!
> Thy crown does sear mine eyeballs. And thy hair,

> Thou other gold-bound brow, is like the first.
> A third is like the former. Filthy hags,
> Why do you show me this?—A fourth? Start, eyes!
> What, will the line stretch out to th' crack of doom?
> Another yet? A seventh? I'll see no more.
> And yet the eight appears, who bears a glass
>   Which shows me many more;

I've never been satisfied with the way this is explained or staged. You could imagine that the last king in line holds the mirror up behind himself, giving us a view of eight kings imaged by the mirror so that they seem to be sixteen. But you'd have to move the mirror so that the effect could be seen by the whole theater, or else why bother with the mirror at all? How would a mirror show an infinite line of kings, anyway?

But if that mirror was accompanied by another mirror in front, and the whole setup moved to show to the whole "house," you could show an infinite line between almost-parallel mirrors. Perhaps some magician or showman was exhibiting a life-sized "infinity box" or "endless gallery" in London, or had done so shortly before the play's opening. If so, the mirror on stage could serve to remind people of the effect. In either case, Macbeth's lines would then refer to a long line of rulers, stretching off into the dim mists caused by the limitations of Elizabethan mirror reflectivity.

Notes

1. Pliny, *Natural History*, section XXXVI.26.66. Another example of this idea was played not for humor, but as horror. In his short story "Midnight in the Mirror World" (originally published in *Fantastic Stories of Imagination*, October 1964, and reprinted in the collection *Night Monsters* [Ace Books, 1969]), Fritz Leiber writes of a man who looks at midnight into the infinite series of reflections between mirrors hung on opposing walls. The eighth reflection shows not just him but also a figure strangling him. On successive nights, the aberrant reflection keeps getting closer.
2. Pliny, *Natural History*, section XXXVI.26.66.
3. Named after the supposed author of a Celtic myth cycle. The works of Ossian, the "Homer of the North," were found to be actually written by James MacPherson, his supposed editor and translator. Before the fraud was found out, the poems of Ossian had inspired several works of music and art, and made the names *Malvina*, *Selma*, and *Oscar* popular.
4. He also gives the first known explanation of the mirror-based illusion called "Pepper's Ghost."
5. https://www.folger.edu/shakespeare-unlimited/magic-teller (accessed November 9, 2019).
6. Act 4, scene 1, lines 108–122.

33

The Great 19th-Century Green Spectacle Craze

"Welcome, welcome, Moses! Well, my boy, what have you brought us from the fair?"

"I have brought you myself," cried Moses, with a sly look, and resting the box on the dresser.

"Ay, Moses," cried my wife, "that we know; but where is the horse?"

"I have sold him," cried Moses," for three pounds five shillings and twopence."

"Well done, my good boy," returned she. . . . Come, let us have it then."

"I have brought back no money," cried Moses again. "I have laid it all out in a bargain, and here it is," pulling out a bundle from his breast.

"Here they are; a gross of green spectacles, with silver rims and shagreen cases."

"A gross of spectacles!" repeated my wife in a faint voice. "And you have parted with the Colt, and brought us back nothing but a gross of green paltry spectacles!"

—Oliver Goldsmith, *The Vicar of Wakefield* (1766)

Moses Primrose, judging from Mrs. Primrose's reaction, has clearly made the worst deal at the fair since Jack sold his cow for a handful of magic beans. The story is appropriate for this article, however. The book was immensely popular with Victorian audiences, and near the beginning, it provides us with a bag full of green spectacles, something the 19th century itself did as well.

This was pointed out by Hiroku Washizu in the Fall 2011 issue of *The Edgar Allen Poe Review*.[1] Poe's detective C. August Dupin wears green spectacles in *The Purloined Letter* (1844) and in *The Mystery of Marie Roget* (1843), using them to hide the movements of his eyes (and his dozing). In *Bon-Bon (The Bargain Lost)* (1832), the devil wears green spectacles, using them to cover the blank spaces where his eyes should be. But Poe is also supposed to have written (in an anonymous article attributed to him) that green spectacles were an abomination.

Poe was not alone in his describing characters with such eyewear. Nathaniel Hawthorne makes note of it on numerous occasions. In the novel *Fanshawe* (1828), he observed that the highest class of scholars wore them, but he noted in his *American Notes* (1835–1853) that the bumpkin did, too, trying to imitate the habits of the elite. In *A Wonder Book* (1851), he states that one character wears them "less for the protection of his eyes than for the dignity that they imparted to his countenance."

Charles Dickens, in *The Pickwick Papers* (1837), describes a "studious-looking" man wearing green spectacles, just as Hawthorne does. Beau Brummel, that avant-garde of fashion, in a novel published in 1844, shortly after his death, was described as wearing green spectacles. In Washington Irving's *The Devil and Tom Walker* (1824), Walker wears green spectacles. *The Class-Book of Anatomy* (1837) by Jerome Smith derides "the ridiculous mania" of wearing green spectacles, "which boys of a certain class seem to imagine adds wonderfully to the dignity of their appearance," and condemns them as injurious to good vision. There are a great many other references to then by now-forgotten authors, in both fiction and nonfiction from the first half of the 19th century. Washizu points to examples in American magazines such as *Atlantic Monthly*, *Harper's*, and *The New American Review*, but there are many examples from the British press as well.

Furthermore, famous individuals notably sported green glasses as well. Thomas Jefferson (1743–1826) owned a pair, now in the museum at Monticello. William Wordsworth (1770–1850) wore them, as attested by many people. The French revolutionary Maximilien Robespierre (1758–1794) was thought to own a pair. Napoleon is depicted in an 1803 cartoon wearing them. Reverend Bennet Tyler (1783–1858), who served as the president of Dartmouth College between 1822 and 1828, had a set that is still in the museum at Dartmouth. I myself have seen a pair of strikingly bright green glasses supposed to have been worn by the Mormon leader Brigham Young (1801–1877). It's not known if Samuel Johnson (1701–1784) wore them, but he discussed them with his biographer, James Boswell.

Is this a real phenomenon? The use of specifically green spectacles seems to overshadow other colors. The frequency of the use of "green spectacles" can be plotted using the Google N-gram viewer. From this we can see that the term was in use by 1800, and its frequency rose to a peak in the late 1830s before dying away briefly, then rising again. If you perform a similar search for the term "blue spectacles" (or similar terms using other colors), you don't find similar usage. Smith, cited earlier, definitely called it a "mania."

What were the reasons for the use? As we've already seen, some people used them to hide the motions of their eyes, or to appear more sophisticated (the "cool" appeal of dark glasses is apparently nothing new). Despite what some have written, sunglasses were *not* invented in the 20th century, and several in the early 19th century wrote of using the dark glasses as protection from the bright rays of the sun, especially in places where there was much bright reflection, such as from mountain snow.

Certainly they had novelty value. *The Nautical Magazine* in 1833 reported a case of sailors amazing island people with them.[2] "A party of them were squatted on the ground, and we had been entertaining them with the green spectacles which some of us wore. They had enjoyed looking through them, and were in ecstasies at seeing every thing appear of a green colour."

Why should this craze for green spectacles, whether because they gave relief to the eye in protecting it from harsh colors, or from bright light, because they hid one's eyes, or gave an aura of mystery and stylishness, erupt in the first half of the 19th century?

Some have pointed to the work of London optician James Ayscough, who in the 1750s advocated the use of green- and blue-tinted spectacles.[3] He felt that these would relieve the strain of harsh white light. He also held that tinted glass was stronger and of better quality. Nevertheless, I find no evidence that Ayscough's theories were influential in steering people 50 or more years later into using green spectacles.

Certainly the use cannot be because green glass or green spectacles had not been available earlier. The emperor Nero famously used to watch events through his *Smaragdus*, a green stone that was probably an emerald. Green glass has been available since antiquity, and green-colored stained glass was used during the Middle Ages. Samuel Pepys rather famously describes buying a set of green glasses in his diary in 1666 to alleviate the strain on his eyes.

The answer, as a historian I spoke with pointed out, might be as simple as the fact that this was the beginning of the Industrial Revolution, which provided inexpensive consumer goods in bulk. Samuel Pepys's green glasses were meticulously crafted by hand, but Moses Primrose was able to obtain a gross of such spectacles for the price of a colt because manufacturing prices had come down, with the new industrialization.

Not only did industrialization lead to cheap spectacles, it led to other inexpensive consumer goods as well, one of these being inexpensive books. Victorians were able to read about Moses Primrose and the gross of green spectacles because the book was inexpensively printed and widely available. And so were the magazines that printed Poe's stories, and that printed Dickens's *Pickwick Papers*, and the works of Nathaniel Hawthorne and Washington Irving, and the numerous lesser-known writers who mentioned green spectacles in their works. Green spectacles may have become a craze because steampower provided the means to make the spectacles in bulk and to widely spread the word about them.

Notes

1. Hiroko Washizu, "The Optics of Green Spectacles," *The Edgar Allan Poe Review* 12, no. 2 (Fall 2011): 48–57.
2. Abrolhos Banks, "The Bijooga Indians," *Nautical Magazine for 1833*, p. 207. https://books.google.com/books?id=D-I1U7X8JiwC&pg=PA207&dq=%22Green+Spectacles%22&hl=en&sa=X&ved=0ahUKEwis5O6E_YnUAhVLRCYKHbbOBps4MhDoAQhNMAg#v=onepage&q=%22Green%20Spectacles%22&f=false
3. "Ray-Ban's Predecessor? A Brief History of Sunglasses," on the webpage *The Chirurgeon's Apprentice*, June 21 2013, http://bowshot6.rssing.com/chan-4388055/all_p3.html (accessed November 8, 2019).

34

Tanagra Theater and the Fishbowl Mermaid

In the 1860s, a farmer's plow in the Greek village of Vratsi turned up a tiny terracotta figure of an exquisitely detailed human being. Within a few years, a great many more were found in the region of Tanagra in Boeotia, and they became immensely popular.[1] Copies of them sold in large numbers throughout Europe and the United States, like Hummel figures today. "Tanagra figure" became synonymous with "tiny and beautiful," so that authors Oscar Wilde and Edward Frederic Benson could describe a petite and beautiful woman succinctly in those terms.[2]

So when around 1910 a new illusion featuring miniature performers on a miniature stage appeared, it was termed the Tanagra theater. The performers were apparently human actors but appeared to have been reduced to a foot or so in height. Tanagra theaters surfaced as attractions across the continent. Some sources say that it originated in Germany,[3] but someone named Sollé was said to have patented it in 1914 in France.[4] In 1923, in Berlin, Friedrich Keisler used such a Tanagra theater set in the center of an ornate mechanical set as part of his production of Karel Ćapek's science fiction play *R.U.R.* (*Rossum's Universal Robots*), the work that introduced the word "robot" to the world.[5] Theatergoers could see apparently miniature actors in the "box," as if watching a television set and then see them full-sized seconds later in front of that set. Edward P. Schreyer of the Tanagra Theater Company brought the idea across the Atlantic, and in 1922 the same illusion was used in New York City to put on a miniature fashion show at the Bijou Dress Company on Fifth Avenue.[6] More recently, in 2014 a "technoillusionist" produced a Tanagra theater biography of Nikola Tesla.[7] The miniature people in all of these appear to be without optical aberrations and appear as if they are three-dimensional figures that could be viewed over a wide range of angles.

It was an optical trick, of course, but most people were unaware of exactly how it was done, content to attribute it to an attitude of "it's all done with mirrors." But we in the optics community, of course, do care about how it was done, and most of you reading this piece have undoubtedly already imagined one or more ways to pull off this sort of illusion. And, in fact, the illusion has been performed in several different ways. The concept of using concave and convex mirrors goes back at least to the 1st century CE, when Seneca wrote about them in his *Natural Questions*,[8] and possibly a great deal earlier. Various illusions presented apparently reduced and moving human figures, such as *The Bottle Imp* exhibited in 1875 at the Royal Polytechnic Institute in London.[9] The Tanagra theater, however, presented a uniquely perfect and sharp image. Think of this as a problem in optical engineering—what is the best way to implement the illusion, in the space available and for the available cost, and producing an image minimally affected by aberration?

For the theater and the fashion show, it was necessary for the image to be erect so that the actors and models could walk around and interact normally. The obvious way to do this is to image real human beings on a brightly lit stage so that the image is reduced, and to use some method to erect the image.

One could, instead, use a large negative lens for viewing the actors, but then one would have to be looking through a large negative lens, which is not only more difficult to come by but is not as effective an illusion.[10]

Diagrams published by the Deutsches Museum in Munich (which has had a Tanagra theater on display since before World War II) show that their version uses a large concave mirror, rather than a positive lens, to accomplish the imaging. It is easier to obtain good large mirrors than lenses, and one need not worry about chromatic aberration. The image is flipped right side up by first reflecting the rays from the subjects using two large flat mirrors set at 90 degrees to each other. Alternatively, one can use three plane mirrors to form a corner cube, which would correct not only up-down inversion but left-right as well, but it's not necessary for most uses of the illusion, and written descriptions list only two mirrors. Using only reflective optics of good quality, this method will give very good images that appear to "float" in space in front of the optics, as long as the concave mirror is in the line of sight. It requires a large concave mirror with a long enough separation between the mirror and the subjects.

This method wasn't only used for miniature theater and advertising. In 1931, in San Francisco, Monk Young opened the #65 Club, a speakeasy and gambling establishment. It was managed by Agostino "Bimbo" Giuntoli, who later took over ownership of the club, which became known as Bimbo's Club 365. The club, which still operates, became famous for its entertainers and for the gimmick of having a fish tank at the bar with a miniature nude mermaid in it. The mermaid, called "Delfina" ("Dolphin"), featured prominently in the bar's décor and advertising and was in reality created by another manifestation of the Tanagra theater, very likely using the arrangement thus described. The model portraying Delfina was actually on a dry stage with an undersea background and was "projected" into the tank by the optics.

The Club 365 mermaid inspired similar images on television and the movies. In the 1991 television movie *Columbo and the Death of a Rock Star*, the titular detective interviews a (non-nude) mermaid "on the job" in her studio under the bar. In this case, the model is envisioned as being suspended on ropes, which the Club 365 mermaid is not. Bobby Jo Moore appears as Mona the Advice Mermaid in several episodes of the Showtime series *The Chris Isaak Show*. And, although the mermaid depicted is not miniature, it seems likely that the mermaid shown in a tank in the wall between the manager's office and the ballroom at the fictional South Seas Club in the 1992 film *The Rocketeer* also owes a debt to the long-running Club 365 attraction.

Advertisers, theaters, and trendy bars can afford good optical elements and a great deal of space. But a travelling carnival can't afford either expensive optics or the space to realize a high-tech illusion. Their equipment must be inexpensive, compact, and highly portable. So it's surprising that the "Girl in a Fishbowl" was a popular attraction at road shows and carnivals, alongside the freaks, the magician, and the Electrified Lady.[11]

One of the places that the carnivals got their ideas from was "Brill's Bible,"[12] a compendium of tricks, illusions, and gimmicks sold to those in the trade by Aaron K. Brill in the years 1954–1976 for a surprisingly low price. Copies today go for large sums, but here you can find budget plans to add some sparkle to your sideshow.

Brill's plan for the "Girl in the Fishbowl" illusion was advertised as using "No Lenses!," a plus for cash-strapped operations that couldn't afford expensive pieces of carefully cut and polished glass. Furthermore, the space for the illusion was very small—it fit into a box no more than three or four feet tall, with the fishbowl on the top.

So how did it work? Examination of the plans shows that there is a single large plane mirror, used to redirect the optical axis of the system from horizontal to vertical. The "mermaid" lay on the bottom of the box housing the illusion, with her head toward the fishbowl/customer end. She would appear, in the final image, to be sitting or floating upright, so the floor of the box was decorated to look like a horizontal look into the ocean.

No lenses were needed, because the mostly spherical fishbowl, filled with water, or possibly clear mineral oil, acted as a lens. It was the ideal solution—the fishbowl efficiently acted as both nominal stage and as the optical element. It was inexpensive, and of sufficiently good quality to produce a good image. Any errors or distortions could be explained as imperfections in the bowl itself. The fluid-filled bowl acted as a spherical lens and produced an ordinary real, inverted image, but the subject was already "upside down" to the folded line of sight, so no erection optics were necessary. With a light bulb or two inside the box (and, one hopes, a fan to relieve the poor woman stuck in the box all day) to illuminate her, the illusion would be perfect, provided the customers were kept far enough back so that they didn't realize that the image wasn't actually *in* the bowl, but floating in space in front of it.

However, this was relatively simple to fix. Patent #4,094,501 was granted on June 13, 1978, to Edward D. Burnett of Chicago, Illinois. Burnett was a magician (his name appears in several newspapers in the 1950s and 1960s. As with many magicians, he seems to have had a sense of humor—some of the "reference patents" he cites seem to have no relationship to his idea whatsoever), and his small addition to the Brill's Bible design makes it much more satisfying. He simply placed another fishbowl in front of the first one, so that the image ended up being projected into the second fishbowl. The first could be hidden behind cowling so that only the second was visible (Figure 34.1).

I constructed both of these illusions on my dining room table, using fishbowls purchased at a thrift shop, and (in place of a nude woman reflected from a mirror) a stuffed plush unicorn turned upside down on a black cloth, illumined with a flashlight. The One Fishbowl Brill method produced a good illusion, but the Two Fishbowl method gives superior results. Not only does the image appear to be inside the second fishbowl, but positioning the second bowl allows you to "fine tune" the image, adjusting the magnification and making small corrections to the quality of the image. Considered in terms of paraxial optics, placing the image produced by the first fishbowl near the center of the second is like positioning an image at the position of the equivalent thin lens representing the second fishbowl. If it coincides exactly, there is no net effect, and the image is the same size and location as if the second fishbowl were not there at all. If the position is off by a small amount, the image will be slightly different in magnification and location, but it's a very small effect.

The optical pioneer Sir David Brewster (1781–1868) is probably best known to most in optics from his work on polarization, as well as the angle named for him. Those working with crystals know about his discovery of biaxial crystals and double refraction. But he also invented the kaleidoscope and the stereoscope, and he was

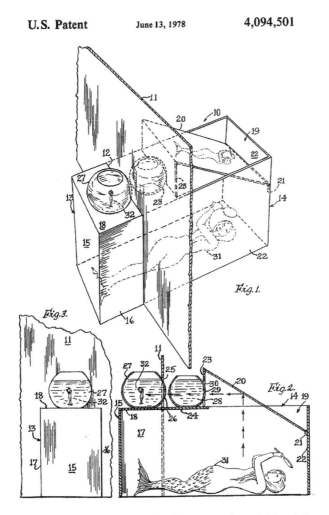

Figure 34.1 Illustration from 1978 patent for the mermaid in a fishbowl illusion.
US Patent Office

much better known in his lifetime for these inventions and his popularization of science. His book *Letters on Natural Magic* (1834, and dedicated to his friend, Sir Walter Scott) sets out the methods used by magicians to perform illusions using optics. He doesn't quite describe the Tanagra theater, but he does tell of producing a miniature image with a concave mirror, and of erecting the upside-down image using mirrors. To him should go the last word in describing this sort of illusion:

> Those who have studied the effects of concave mirrors of a small size, and without the precautions necessary to ensure deception, cannot form any idea of the magical effect produced by this class of optical apparitions. When the instruments of illusion

are themselves concealed, —when all extraneous lights but those which illuminate the real object are excluded, —when the mirrors are large and well polished and truly formed, —the effect of the representation on ignorant minds is altogether overpowering, while even those who know the deception, and perfectly understand its principles, are not a little surprised by the effects.[13]

Notes

1. Wikipedia article on "Tanagra figures." https://en.wikipedia.org/wiki/Tanagra_figurine
2. Oscar Wilde in *The Picture of Dorian Gray* and the play *The Ideal Husband* and Edward Frederic Benson in his novels *Margery* and *Juggernaut.*
3. Joe Nickell, *Secrets of the Sideshows* (Lexington: University of Kentucky Press, 2005), 291–293.
4. Sam H. Sharpe, *Conjurer's Optical Secrets* (Calgary, Alberta: Hades, 1985, 1992), 142, 161–163. Sharpe says that the show ran in Paris for several months, and that he has no earlier record of the illusion. He says that he has no record of the patent, and I have been unable to locate it through a patent search.
5. http://scan.net.au/scan/journal/display.php?journal_id=34
6. *Hide and Leather: The International Weekly Shoe Factories—Tanneries—Allied Industries* 64, no. 10 (September 22, 1922), 46. https://books.google.com/books?id=4Zc7AQAAMAAJ&pg=RA9-PA46&dq=tanagra+theater&hl=en&sa=X&ved=0ahUKEwib6_vf4_rLAhVS72MKHXxTAaIQ6AEIHTAA#v=onepage&q=tanagra%20theater&f=false
7. http://www.openculture.com/2014/10/the-electric-rise-and-fall-of-nikola-tesla.html
8. Lucius Annaeus Seneca, *Natural Questions*, section I.16, 82–89, in the Loeb edition, trans. Thomas H. Corcoran (Cambridge, MA: Harvard University Press, 1971, 1999).
9. Sharpe, *Conjurer's Optical Secrets*, 160–161.
10. Nevertheless, the effect has been performed this way. It is described in US Patent #2,232,110 dated February 18, 1941, issued to Alfred Gruenhut, assigned to Yermie Stern Productions. They are on record in *The Film Daily* and *The Motion Picture Daily* in 1938 as using such a mechanism to create the illusion of a living, breathing Santa Claus three inches tall, walking around his workshop. See https://archive.org/stream/filmdail74wids/filmdail74wids_djvu.txt. The use of concave lenses is also described by S. H. Sharpe in *Conjurer's Optical Secrets* and by David Brewster in *Letters on Natural Magic* (see note 13).
11. Joe Nickell, *Secrets of the Sideshows* (Lexington: University of Kentucky Press, 2005), 291–293.
12. Properly the title is *A Brill's Bible of Building Plans and Collection of Much Information Useful to Showmen, Carnies, Fairmen, and Amusement Park Operators*, published by A.B. Enterprises of Peoria, Illinois, from the 1950s through the 1970s.
13. David Brewster, *Letters on Natural Magic* (1834). It was dedicated to his friend, the author Sir Walter Scott. The quotation comes from Letter #4, on p. 68 of the edition published by Harper and Brothers of New York in 1843. https://books.google.com/books?id=yf2oGn1oS7YC&printsec=frontcover&dq=%22Letters+on+Natural+Magic%22+Brewster&hl=en&sa=X&ved=0ahUKEwjb1tiCrbDMAhWLph4KHfDfA9gQ6AEIJjAA#v=onepage&q=%22Letters%20on%20Natural%20Magic%22%20Brewster&f=false

35
The Secrets of X-Ray Spex

In the 1960s and 1970s, the advertisements for X-Ray Spex appeared on the inside covers of comic books. They still sell them at novelty stores and on the Internet. The cheap, black-rimmed cardboard spectacles are made with cardboard filling all but a small round hole in the center, the rest being taken up by alternating red and black concentric circles, creating a "hypnotic eye" effect. They're X-Ray Spex, the ludicrously inexpensive devices that promised to show you the bones in your hand, or possibly the body of another person beneath their clothes. These mysterious devices have been either amazing people or disappointing them since 1906, when the first ones were patented in the United States. Similar devices were patented in 1909, 1950, and then really burst upon the market in 1964. At that time, they were widely marketed by novelty dealer Harald Nathan Braunhut, who also "invented" sea monkeys (he, too, obtained a patent for them in 1971).

We'll dispense with the first "mystery" right off the bat. Braunhut's x-ray glasses, like virtually all of them before his, worked by having a piece of carefully oriented almost transparent bird feather inserted into each of the small eye holes. The feather acted like a very inexpensive diffraction grating. The barbs on bird feathers are typically on the order of 100 microns (0.1 mm) apart, corresponding to 10 line pairs per millimeter, a pretty coarse grating. Five hundred nanometers of light would have the separation between orders at about one-third of a degree, and with little separation between colors. The result is that the lowest orders overlap with a slight offset, creating a dark shadow surrounded by a lighter gray shadow. The dark shadow is interpreted as the bones in the fingers, or the contours of a body beneath the clothes. A very innocent reality underlying the risqué premise. Braunhut knew how to package it for maximum impact—show the wearer apparently observing the bones in his hand while a lady stood some distance off. The device sold.

Braunhut (who called himself "von Braunhut" and reportedly flirted with neo-Nazism) continued to sell the original feather X-Ray Spex until he stumbled across another pair of novelty glasses manufactured by American Paper Optics, which had diffractive optical element (DOE) "lenses" that produced images of hearts around a bright source. Braunhut found that they also produced the same "x-ray" effect as his devices, and contacted APO, claiming that they plagiarized his device. APO said that they weren't trying to infringe on his territory—the x-ray effect was a fortuitous (or in-fortuitous) side effect of their diffractive glasses. They reached an agreement where they would provide Braunhut with DOE lenses to use in place of his feathers—the DOEs gave a much clearer and cleaner effect—in exchange for no legal issues and a share of the profits. The optical element they came up with produced the words "X-Ray" if you look at a bright point source. (You can also shine a laser pointer through one of the lenses and project the words on a wall). This allowed them to say "Amazing X-Ray Vision Guaranteed!" on the packaging, since you could, indeed, see the words

"X-Ray." Guaranteed. (At the same time, a Hong Kong company, Ming Shing Plastic and Metal, continues to produce a feather-based pair of "X-Ray Gogs" that look almost identical.) Harold died over 15 years ago, but, as of this writing, his widow still collects a small royalty from the device.

A little convoluted, but understandable enough. But I have two questions yet. The first one is, who figured out that a bird's feather could act as a low-cost acceptable quality diffraction grating? That light passing through a feather becomes separated into colors has been known since at least the time of Robert Boyle in the 1660s. James Gregory appears to have independently observed the same thing in 1670. Internet articles hail Gregory as the "discoverer of the diffraction grating," but it appears, rather, that he observed a physical phenomenon, and there is no sign that he could explain it (see my article "Light ... as a Feather" in the February 2014 issue of OPN and Chapter 8 in the present volume). Gregory corresponded with Newton regarding this, but Newton's commitment to the corpuscular theory prevented him from explaining the effect in terms of interference.

Thomas Young, that champion of the wave theory of light, looked for examples of interference everywhere, including the double-slit experiment, supernumerary rainbows, and a diffraction grating in the form of a finely ruled instrument scale, but he seems to have missed the interference in bird feathers.

The first proper recognition of the effect of bird feathers acting as a diffraction grating and causing separation of colors that I could locate appears to be in a letter by "J. K." of Belfast in the *Belfast Monthly Magazine* for March 20, 1812. In "Solution of an Optical Experiment," he (or possibly she) refers to a letter in the previous month's issue, in which "J. S." of Armaugh, repeating an experiment by Dr. John Simpson, observed that passing light through a feather caused it to break into colors. J. S. thought that the oil in the feather might be acting as prisms.

J. K., on the other hand, observed that each interstice acted as a tiny aperture that was small enough to cause the light to break into different colors, and that the color rays from each tiny aperture added in concert with those of the other apertures. Certainly, within a couple of decades, it was well known that feathers acted as diffraction gratings; there are many references throughout the 19th century. It would have been well known by the time people started getting patents on their various incarnations of x-ray glasses.

But here is the last mystery associated with this device—Braunhut in his 1971 patent explicitly calls for the use of feathers. So does C. S. Raizen in his 1950 patent, and Fred J. Weidenbeck in his 1909 patent. But George W. MacDonald, in his 1906 patent (which appears to be the first), never specifies using feathers as low-tech, low-cost gratings. Instead, he calls for a "transparent or translucent medium ... provided with fine parallel lines *a*, produced by cutting, etching, pressing, or other operation ... the transparent or translucent medium may be of glass, isinglass, gelatin, or other suitable substance."

What could MacDonald have had in mind? Ruled gratings were expensive at the time, and they were time-consuming to make. It would make for a very expensive novelty if each grating had to be separately manufactured with a ruling engine. To be inexpensive enough to sell, the gratings must be mass-produced. They must be replica gratings.

Arguably the first replica grating was produced accidentally when David Brewster was studying the colors of mother-of-pearl. He waxed his specimen in place, and upon breaking it off, found that the surface of the way, which duplicated the fine lines of the original, produced the same colors. In the mid-19th century, there were photographically reproduced gratings. But around the turn of the century, several astronomers were producing replicas by using various mixtures spread on a master grating. Robert James Wallace, F. E. Ives, and Thomas Thorp used various mixtures to obtain their replicas. Thorp's method consisted of using celluloid in solution poured onto a grating that had been oiled. After it had set, the celluloid film was peeled off and cemented to a glass plate with a mixture of gelatin and glycerin. Wallace criticized it as producing too many air bubbles that scattered light out of useful orders. But such a grating would have been more than adequate for MacDonald's novelties, and still far better than feathers. Perhaps he planned on purchasing rejected replicas, or he envisioned making his own using not proper gratings but quasi-gratings such as "Barton's buttons."

Whatever the case, it's interesting that, after a century or so of being made with bird feathers, X-Ray Spex are now being made with diffractive optical elements, as MacDonald originally envisioned them.

I owe special thanks to John Jetit, president of American Paper Optics (the current manufacturer of X-Ray Spex), and to Joseph Vandergracht of Holospex (which makes the DOEs for American Paper Optics) for discussions and for sending me samples.

References

Boyle, Robert. *Experiments and Considerations Touching Colour* (1670) and *The Works of the Honourable Robert Boyle in Six Volumes.* Vol. 1, Experiment 19 (1772), p. 743.

Demarais, Kirk. *Mail Order Mysteries: Real Stuff from Old Comic Book Ads.* Richmond, CA: Insight Editions, 2011.

Gregory, James . "Letter to John Collins." May 15, 1673.

"J. K." letter. *The Belfast Monthly Magazine* 8, no. 46 (May 1812): 361–365.

Parker, A. R., et al. "A Goniometric Microscope to Quantify Feather Structure, Wettability, and Resistance to Liquid Penetration." *Journal of The Royal Society Interface* 11, no. 96 (2014): 1–30

Wallace, Robert James. "Diffraction Grating Replicas." *The Astrophysical Journal* 22 (1905): 123.

Patents

US Patent 839016
US Patent 914904
US Patent 2527332
US Patent 3592533

36

Fluorescent Paint before Day-Glo

Fluorescent paints, according to the Internet and several books, were invented by the brothers Robert and Joseph Switzer, who were living in Berkeley, California. It was the summer of 1933, and Bob, unloading cartons of tomatoes from a freight car at the H.J. Heinz research lab, fell and seriously injured his head. His vision was permanently impaired, and he had to remain in darkened conditions as he recuperated. During this time, the brothers investigated fluorescent materials using an ultraviolet lamp. They identified several candidates and experimented with mixing these pigments into fluorescent paints, finding combinations that would allow the materials to fluoresce brightly.

One use they found for these paints was stage magic. Robert demonstrated the Magic Balinese Illusion, making a fluorescent-painted dancer seem to appear and disappear under black light. This won an award at a magicians convention in Oakland, California, in 1934. The brothers formed the Fluor-S-Art company and produced black light displays for advertising.

They partnered with Warner Brothers, producing eye-catching posters. They improved the production of fluorescent paints, eventually specializing in the types of pigments that would absorb ultraviolet from sunlight and re-emit it as dazzling visible colors. The technology was used by the military in World War II, and afterward it was grabbed by industries to use for eye-catching packaging. The brothers ended their association with the Warner's subsidiary and founded Switzer Brothers in Cleveland, later the Day-Glo company. The brothers continued to operate the company until 1985, when they sold it. With the proceeds, Robert set up the Robert and Patricia Switzer Foundation, giving scholarships to students (as Switzer himself had received a Scaife Foundation scholarship).

The Switzer story is told in many places.[1] In 2009, it was the subject of a children's book, *The Day-Glo Brothers* by Chris Barton. On September 8, 2012, the Day-Glo Cleveland factory was designated a National Historic Chemical Landmark by the American Chemical Society.[2] If you look in these places, or in *The Encyclopedia of Bad Taste*, or on *Wikipedia*, you will find that the Switzer brothers invented black light paint.

But did they?

Not to detract from the hard work, determination, and successes of the Switzer brothers, but it would seem odd that for a phenomenon of such long standing as fluorescence, and the knowledge of ultraviolet light going back to Johan Wilhelm Ritter in 1801, it would have taken as long as 130 years before someone thought of using this as paint.

The existence in nature of brighter colors than people could achieve with most pigments has been known for a long time—flowers, bird feathers, the colors of tropical fish. Even more intriguing were cases where some object appeared to glow with a light different from that of the overall color. This was observed in the 16th century for the case of extracts from some tropical woods. In the early 19th century, a similar phenomenon was observed with the minerals called *fluorite*, and soon David Brewster observed the effect in chlorophyll, and John Herschel in solutions of quinine.

At first, these effects were thought to be some sort of atypical scattering effect. It was George Gabriel Stokes who performed a series of experiments illuminating such substances with isolated ultraviolet light and showed that this irradiation with invisible light could produce a visible emission that provided the final piece of the puzzle. Stokes named the effect *fluorescence* after the fluorite that exhibited the effect.[3] (It was years later that the shifting of emission to longer wavelengths than the exciting light was dubbed the "Stokes shift" in his honor, and well after that until its cause was discovered in the form of quantum mechanical energy levels.)

Fluorescence proved to be a useful laboratory tool, but the effect was largely a curiosity as long as there was no way to produce efficient sources of ultraviolet light. As Stokes himself observed, candle light had too little of it to provide a visible glow. Sunlight had more ultraviolet than most combustible sources. It was later found that arc lights, discharge tubes, and burning magnesium metal could produce ultraviolet emission, and that one way to eliminate the longer visible wavelengths was to use a deep blue or violet glass filter.

All of this was in place by the end of the 19th century, and it is then that we would naively expect ultraviolet paint or ink to make its first appearance. In fact, biologists doing fluorescence studies on animal and vegetable substances frequently called these "fluorescent pigments," which ought to have excited someone's interest.

An 1882 book, *Light: A Course in Experimental Optics* by Lewis Wright, suggests using the light from burning magnesium passed through blue filters to isolate ultraviolet light, and it suggests numerous substances which one can use to observe this effect—petroleum (which fluoresces blue), tincture of turmeric (green), uranium nitrate or uranyl glass (yellow), and thallene, a petroleum extract prepared by Prof. Henry Jackson Morton of the Franklin Institute. It fluoresced with a very strong green (which prompted the compound's name, from the Greek word for *green*).[4] Significantly, Wright observes that "designs painted with it show up brilliantly in almost invisible violet light."

Clearly the idea of producing a fluorescent paint had been conceived of, even if this is a long way from a commercial product. Later works observed the fluorescence of inks containing *eosine* (which is not a single substance, but actually a collection of related dye compounds. They were discovered by German chemist Heinrich Caro. The name was derived from the nickname of his childhood friend). In 1911, E. Arthur Bearder presented a paper (later published) that listed a couple of dozen organic dyes and substances known to fluoresce, and speculating upon the molecular structures responsible for such behavior.[5] Several writers made note of *Balmain's Paint*, a phosphorescent paint that fluoresced under ultraviolet light. Its active ingredient was calcium sulfide. It was patented in the United States by William H. Balmain.[6]

The ability of certain substances to produce brighter colors by fluorescence was recognized by 1903. Examination of articles from the turn of the century turns up several about fluorescent brighteners. The ability of substances to absorb ultraviolet light and re-emit it in the visible was essential to Peter Cooper Hewitt's invention of the

fluorescent tube, and he mentions them in his many patents.[7] Some come close to suggesting fluorescent paint, and one observes that certain binders will inhibit fluorescence. Indeed, the Cooper Hewitt Electric Co. was the assignee for Leroy J. Buttolph's
patent for "fluorescent paint."[8] Buttolph was an engineer who made several contributions to fluorescent lighting, but he here seems to have moved into different territory.
He explicitly states that the paint is "for use on theatrical costumes and scenery and in
other displays wherein the parts painted are subjected to ultra violet radiation for the
activation of such fluorescent material and its simultaneous display to spectators in
the dark to produce fluorescent effects," thereby anticipating the Switzer Brothers by
several years. He gives several formulations, but the only fluorescent material he lists
is anthracene.

A similar patent was filed on August 9, 1922, by Nathan Sulzberger.[9] Titled
"Apparatus and Method for Producing Novel Lighting Effects," it suggests more candidates than anthracene alone, listing "Fluorescein, Anthracene, Uranine, Rhodamine,
etc." The patent furthermore suggests the use of sodium silicate (which is very transparent in the ultraviolet) as an adhesive. It notes that the color of the fluorescence may
be different from its apparent color in ordinary light, and the proposed use is for advertising signs, illuminated by ultraviolet light (possibly from a mercury lamp filtered
through Wood's glass).[10]

A 1920 article speaks of zinc sulfide fluorescent paint being used for dials by the
military, glowing under ultraviolet light. A 1920 reference book mentions one of
the uses of cadmium tungstate as "fluorescent paint." In 1927, J. Broadhurst wrote of
painting fences and bridges with fluorescent paint. It's likely that some fluorescent
inks and paints were used for transmitting secret messages as a form of "invisible ink"
during the first part of the 20th century, especially during World War I.

The Switzer brothers very likely were unaware for the previous efforts by Sulzberger
and Buttloph and of their respective patents, or of the occasional other uses of fluorescent paints. In the absence of widespread reporting of such esoteric phenomena,
or of as versatile a search mechanism as today's Internet, it would be easy to miss such
developments. Certainly, although people had made fluorescent paint for their own
applications, no one seems to have marketed commercial fluorescent paint. So when
the Switzers began their work, they had a completely open market before them, and
they explored not only fluorescent materials but also how to compound paints and
how to produce a gamut of colors. Their efforts not only produced a range of bright,
ultraviolet-activated pigments but also something new—a line of colors whose fluorescence could be clearly seen under ordinary daylight, making preternaturally bright
colors. The military saw the advantage of such bright colors for visibility. Many years
later, the colors were adopted by corporations to make boxes stand out from their
competition on store shelves. And the bright colors seen under "blacklite" illumination were used in displays, in shows, in museum exhibits (such as the New York
Hayden Planetarium's blacklite corridor, which opened in 1954 and continued until
the original building was replaced in 1997), and ultimately in the 1960s phenomenon
of blacklite posters. The Switzers may not have invented ultraviolet paint, but it was
by their efforts and industry that they expanded its capabilities and made "Day Glo" a
household term.

Notes

1. For instance, on Wikipedia—https://en.wikipedia.org/wiki/Blacklight_paint or the Robert Switzer obituary in the *New York Times*, August 29, 1997.

2. https://www.acs.org/content/acs/en/education/whatischemistry/landmarks/dayglo.html

3. George Gabriel Stokes, "On the Change of Refrangibility of Light," *Philosophical Transactions of the Royal Society of London* 142 (1852): 463–562.

4. Lewis Wright, *Light: A Course of Experimental Optics, chiefly with the Lantern* (New York: MacMillan and Co. 1892), chapter 8, 134–146. Quote on p. 142. https://archive.org/details/lightcourseexper00wrigrich/page/n8.

5. Arthur Bearder, "Fluorescence," *Journal of the Society of Dyers and Colourists (later Coloration Technology)* 27, no. 12 (December 1911): 270–279.

6. William H. Balmain of Eversley, Isle of Wight "Self-Luminous Paint," US Patent 264,918 (September 26, 1882).

7. Most notably in "The Art of Lighting," US Patent 1150118 (filed September 18, 1909, granted August 17, 1915).

8. Leroy J. Buttolph, "Fluorescent Paint," US Patent #1658476 (filed December 17, 1924 and granted February 7, 1928).

9. Nathan Sulzberger, "Apparatus and Method for producing Novel Lighting Effects," US Patent 1,617,425 (Feb 15, 1927).

10. Sulzberger is an interesting individual. He was apparently a member of a meat-packing family. He was a friend of the mystic writer and poet Rainer Marie Rilke, and was himself a lyricist, publishing both under his name and the pseudonym "D. R. Enness," where "Enness" seems inspired by his initials, "N. S." He filed a great many patents on a variety of topics. One of his more interesting ones, in this regard, was meant to prevent overheating of cinema screens, which caused many fires in the early years of movies. His idea was to project an image using only filtered ultraviolet light onto a screen coated with fluorescent material. U.S. Patent 1592393, filed August 12, 1921, and granted July 13, 1926, for "Projecting Apparatus, etc." One has to admire his ingenuity, although I suspect that the difficulty in building projection apparatus and film that will work in ultraviolet light would have been too difficult and expensive to make this worthwhile. On Sulzberger, see Ralph Freedman, *Life of a Poet: Rainer Maria Rilke* (Evanston, IL: Northwestern University Press, 1996).

37

The First 3D Movies

For a long time, 3D movies seemed to come in cycles of about 20 years, although recently they seem to have leaped from a fad to become ubiquitous. But there were outbreaks of 3D movies in the 1990s, the late 1970s, the early 1950s, and even the 1920s.[1] There were even patents filed in the 1890s by Frederick Henry Varley[2] and William Friese-Greene,[3] in which they proposed projecting two images onto a screen in different colors, to be viewed with different colored lenses over each eye, an early effort at anaglyphic 3D. Nevertheless, there was an even earlier burst of activity, which unfortunately did not break into commercial success.

When Thomas Edison created one of the first motion picture devices and wanted to commercialize it, he got his team working not on a projector to display films to a theater filled with patrons, but a peep-show device that he called the "kinetoscope," which let one person at a time pay a coin and squint through a lens for a private show. Chief among the design team was a young inventor of Scottish-American ancestry named William Kennedy-Laurie Dickson (1860–1935). Edison's kinetoscope was displayed in "kinetoscope parlors" that contained many such devices, each with its own short film.

The kinetoscope was expensive, however, and the nitrocellulose film was prone to frequent breaking. Dickson, after a falling out with Edison, went off on his own and invented a competing device in collaboration with another inventor, Herman Casler. This was the "mutoscope," and it solved the problems of expense and frequent breakage by replacing the roll of film with a succession of photographs printed on heavy stock and bound into a cylinder. It was essentially a giant "flipbook," in which the succession of pictures were flipped by rotating the cylinder with a crank handle while a metal "finger" released them one at a time to view under an electric light. A lens was usually placed one focal length from the viewed frame, and a metal frame excluded light.

The new mutoscope was an immediate success, and mutoscope parlors sprung up as the kinetoscope parlors had. Since the mutoscope typically cost a penny to view (far less than the kinetoscope shows), these became the first penny arcades, taking the name from a venue in the Midwest. Many mutoscopes survive to the present day and are curiosities at amusement parks and the like. Many of the original mutoscope reels survive as well.

I had unearthed this history while researching an early American amusement park, and I was at one point struck by a thought. At the same time that Edison's kinetoscope and Dickson's mutoscope were establishing the institution of the penny arcade, another device was continuing its round of popularity. This was the stereoscope. Stereoscopes had been invented and improved by several people between about 1830 and 1860. The most famous design, often mistakenly called today the "stereopticon," was properly called the "Holmes stereoscope." It really was invented by the physician and poet Oliver Wendell Holmes, who did not patent the device. It was simple and elegant and inexpensively made, the most expensive part being the thick main lens that was sawn in half and each half used "backward," with the thin edges facing each other. This not only gave comfortable viewing but also gave the needed prismatic effect to coalesce the two images one for each eye.

Dickson could not have been unaware of the similarity between his own lens-equipped mutoscope and the stereoscope. It didn't seem possible that he could overlook the possibility of combining the two devices to make a viewer for three-dimensional flipbook motion pictures—all he would need to do was to substitute the sawn and reassembled lens for his single viewing lens and produce a series of flipbook cards carrying the two stereo images.

I immediately began a search for the stereomutoscope, or mutostereopticon, or some other variation on the possible names for such a device and found ... nothing. Well, almost nothing. I found some devices painfully close to the concept. The Mills Cathedral Stereopticon Machine was a circa 1910–1920 device that provided the viewer with a series of stereo views on a cylindrical flipbook, viewed through paired lenses—but they were a succession of unrelated "still" 3D images, not a motion picture succession. In the 1950s, there were "3D Art Parade" machines that used paired lenses viewers to look at stereo pairs of color photographs, often of scantily clad ladies. But again, these were a succession of stereo "stills," not a flipbook with the image of motion. The closest thing I found were a modern website describing and showing the construction of a modern stereomutoscope and a kickstarter site devoted to raising funds to build another one. Clearly I was not the only person struck by this idea and its easy construction. But these were very recent incarnations of the idea. I had hoped that I would find evidence of this in use in turn-of-the-century arcades.

It turned out that several people did, indeed, have precisely that idea, as I learned from the book *Stereoscopic Cinema and the Origins of 3-D Film, 1838–1952* by legendary 3D expert and historian Ray Zone. As he points out, a stereomutoscope was patented in Great Britain by Dickson in 1899.[4] He received a US patent for the same device in 1903.[5] Another such device was patented in 1900 by Frank Muniot and Louis Garcin of New York City.[6] Charles Francis Jenkins, an inventor from Atlanta, Georgia, received a patent in 1901 for his "stereoscopic mutoscope," which most closely resembles our notion of what such a device should look like.[7] A stereomutoscope card from 1908 manufactured by the American Mutoscope and Biograph Company—Dickson and Casler's company—was in Zone's possession.

With all these patents, you would think that someone would have marketed such a device—for novelty's sake, if nothing else. But a search through contemporary periodicals reveals that the only citations for anything like this are speculative pieces in technical and scientific magazines. There are no notices in *Billboard* or *The New York Dramatic Mirror* or any other entertainment periodicals that would indicate that anyone had ever put such a device of public display.

Why not? American Mutoscope was tooled up to produce mutoscopes in bulk. It wouldn't take much diversion of resources to make a stereomutoscope. They even made at least one test card, which implies the existence of at least a prototype. But perhaps it was still too expensive and interest in it too low to justify the extra expenditure. As Zone observed, "it is questionable whether the actual device was ever marketed."

The earliest mention of such a device appears in one of Dickson's 1880 note-books. So it appears that he was the first to imagine combining the mutoscope and the stereoscope—not surprisingly, since he also envisioned the mutoscope in the first place. What is interesting is that Dickson also appears to have been the first one to combine sound with motion pictures. An 1889 film of his—now lost—showed him welcoming Edison back from the Paris Exposition, the words recorded on a cylinder. Some five years later he made what is called the "Dickson Experimental Sound Film," which does exist. It shows a man—thought to be Dickson himself—playing a violin into the huge bell of a phonographic recorder while two men dance. The accompanying cylinder has been located and the sound synchronized with the image, making the oldest extant sound movie. It is interesting and intriguing that the man responsible for the earliest sound movie was also responsible for the first 3D movies, and that his name is largely unknown to the general public.

Notes

1. On the very earliest 3D patents, see Ray Zone, *Stereoscopic Cinema and the Origins of 3D Film, 1838–1952* (Lexington: University Press of Kentucky, 2007), 61. https://books.google. com/books?id=UXTAAgAAQBAJ&pg=PA59&lpg=PA59&dq=Ray+Zone+The+Problema tic+Mr.+Greene&source=bl&ots=K_rXjcwD0Y&sig=IWrWpW5N0BS79eCW2gpJxf4pzr g&hl=en&sa=X&ved=0ahUKEwjmu67BtoDNAhUJ0h4KHfSCCCwQ6AEIHzAA#v=one page&q=Ray%20Zone%20The%20Problematic%20Mr.%20Greene&f=false
2. British patent 4704, March 26, 1890.
3. British patent 22954, November 29, 1893.
4. British patent 6794, for a "Stereo Optical System."
5. US patent 731,405.
6. US patent 653,520.
7. US patent 671,111.

38

Perspective Machines, Zograscopes, Megalethoscopes, and *Boites d'Optiques*

In the 18th-century section of the new Art of the Americas wing at Boston's Museum of Fine Arts, there is a very interesting piece. It appears to be a five-foot-tall stubby obelisk of painted wood, with a large cameo-like carving on the front, believed to be the work of woodcarver and architect Samuel McIntire of Salem. If you move aside this carving, you reveal a single large lens, about eight inches in diameter, which is definitely an odd thing to find on a piece of 18th-century furniture. The placard identifies this as a "perspective machine." It is not the device of that name used by Albrecht Dürer and others to properly project an object into a perspective view, but rather a form of 18th-century popular entertainment.

There is a door at the back, which opens to reveal a shelf onto which a print may be placed. The door may be left open to admit illumination onto the print. A 45° mirror at the top of the obelisk directs the optical path from the vertical that extends above the print into the horizontal one centered on the lens. The print is placed at the focal point of the lens, so that a viewer looking at the print sees it effectively at infinity. There is a storage area for more prints.

Why? What was the point of this device, and how was it used? Why not simply look at the print directly, instead of placing it in this optical contraption? The answer is that by looking through the lens, one got a decidedly different viewing experience than by simply looking at the print directly. Viewing it through the large singlet lens, with its aberrations, gave the impression of looking at something on the surface of a somewhat spherical surface that rotated as the viewer moved about from side to side, or up and down. The image had an unusual "liveliness" to it that was completely unlike the static appearance of a flat drawing. Moreover, to many people it appeared as if a street scene so viewed—and the subjects of such viewers were generally panoramic scenes of cities—the scene seemed to take on an appearance of three dimensions, even though this was not what we would call a stereoscopic image. This feature has been the topic of several papers on perception.

These devices might occupy a conspicuous place in the parlor of a well-to-do owner, where its respectable but unusual appearance would invite comment from visitors, giving the owner a chance to display the device. The Museum of Fine Arts obelisk had at one time been the property of Dr. Samuel Holten of Danvers, president of the 1775 Continental Congress and signer of the Articles of Confederation. A similar device had been purchased by Joseph Warren Revere (son of Paul Revere) in London in 1800.

Another sort of perspective machine had the viewing device built into a desk, where it could be hidden away when not in use or could be opened to appear very much like an early television set, with its tiny "screen" of a lens set in a wooden cabinet. One of these is now in the Peabody Essex Museum in Salem, Massachusetts (item

#138369), the work of Edward Johnson of Salem, who constructed at least two such desks between 1793 and 1811 (The other is at the Winterthur Museum in Winterthur, Delaware).

Such devices were also called "zograscopes" or *boites d'optique* (which means simply "optical box"), and they were rare in the Americas. The viewing devices and the panoramic scenes (called *vues d'optique*) were popular for about a century and a half, from 1690 to 1840. They were produced in London, Paris, Basano in Italy, Augsburg in Germany, Amsterdam, and Vienna. In more exotic machines, the prints had cutouts covered with colored tissue and could be illuminated from behind to give "night" views of the cities. The relatively late megalethoscope, patented in Italy about 1865, was especially intended for this use (one such device resides in the collection of the Museum of Science in Boston).

Not all devices were as elaborate and well-made as the ones described here. "Economy" models consisting of little more than a framework, a lens, and a mirror that could be opened out at 45° for use, or folded against the lens for storage, took up less space and cost less. The device was placed on a table, with the print resting at its base. Finally, for those unable to afford such luxuries, there was the traveling peep show, in which an itinerant performer carried his device and pictures from place to place, displaying the pictures along with a verbal description for a fee. Some of the pictures used for these shows could themselves be moved and manipulated.

I am particularly grateful to Dennis Carr for providing many references and for his own thesis on the topic.

References

Ames, A. "The Illusion of Depth from Single Pictures." *Journal of the Optical Society of America* 10 (1925): 137–147.

Balzer, Richard. *Peepshows: A Visual History.* New York: Abrams, 1998.

Blake, Erin C. "Zograscopes, Virtual Reality, and the Mapping of Polite Society in 18th Century England." Paper presented at the Seminar of Cartography, Newberry Library, Chicago, April 29, 1999. http://mitpress.mit.edu/books/chapters/0262572281chap1.pdf

Carr, Dennis Andrew. "Optical Machines, Prints, and Gentility in Early America." Master's thesis, University of Delaware, 1999.

Chaldecott, J. A. "The Zograscope or Optical Diagonal Machine." *Annals of Science* 9, no. 4 (1953): 315–322.

Chaldecott, J. A. "The Zograscope—A Forgotten Name in Optics." *Bulletin of the British Society for the History of Science* 1 (1953): 227–228.

Nagata, S. "How to Reinforce Perception of Depth in Single Two-Dimensional Pictures." NASA, Ames Research Center, Spatial Displays and Spatial Instruments, 1989.

39

The Claude Lorrain Mirror

In the late 18th or early 19th century, you might see something unusual out in the picturesque countryside—a company of aesthetic tourists or amateur artists out to appreciate the scenery and looking for artistic vistas could be seen using an unusual optical device. They would be looking for some ideal or beautiful image, framing it in their device and setting up the best compositions.[1]

> It may be a bit of lake or mountain scenery, or of cloud effect, which now seems too beautiful for reality. Or it is a turn of the country road which the observer is passing; an opening in the woods; a little wayside mill with its rustic surroundings. Perhaps it is only the front yard, with its pretty shrubbery, of his own home; or the rear of his house, with its barn and sheds and meadow lots beyond. Whatever it is, it seems transfigured in that glass.

What were they using to line up these shots, and why? This was, after all, before the invention of photography. It was even before the invention of that forerunner of the amateur snapshot, the camera lucida.[2] These enthusiastic would-be artists had no way to record the images they were industriously lining up and composing. It was as if they were practicing for the eventual coming of the camera.

The device they were using was the Claude Lorrain mirror, which was a convex mirror made of some hard dark substance, like obsidian or jet, ground and polished but not coated, so that the Fresnel reflection alone made it a mirror. Such mirrors have a long history. It is probably a dark convex mirror that a woman looks into to see herself in the woodcut *Vanity* by Albrecht Dürer. It certainly is a dark convex mirror in the case of the woman evidently being punished for the sin of vanity in the "Hell" panel of Hieronymus Bosch's triptych *The Garden of Earthly Delights* (circa 1500). As these images suggest, the convex black mirror had an unhealthy association. They were said to be used by witches for scrying (Figure 39.1).

By the late 18th century, however, it had shed these associations, and small black convex mirrors, fitted into padded cases, were seen as the essential tool for appreciating beautiful scenes. The mirror allowed the user to "frame" his or her composition. The convex surface rendered the image smaller, and the dark coloring reduced the light and, it was claimed, made gradations of color and lighting more evident. It was said to make the image resemble the works of the Baroque artist Claude Gelée (1600–1682), who changed his name to Claude Lorrain. He specialized in large landscapes that featured all levels of lighting within them, from bright sun to deep shadow. There is no other reason to associate his name with the dark mirror, which we have no evidence of him using. Lorrain's name may not be a household name today, but typing his name into the Google N-gram viewer shows that interest in him, as evidenced by references in printed works, peaked around 1820.

PH*337215

Figure 39.1 A Claude Lorrain mirror in its padded case from the collection of the Smithsonian Institution.

I wanted to try a Lorrain out for myself and found that such mirrors aren't easy to come by. I could have one specially ground and polished as a custom job, or I could purchase one from the websites selling modern reproductions, but these were expensive options. I decided to use a less expensive and more easily obtained substitute— I purchased a Magic Eight Ball. The black plastic sides of this mass-produced black plastic sphere function as an acceptable mirror, and it is very nearly spherical, although the radius of curvature is somewhat smaller than that of most Claude Lorrain mirrors.

The mirror does, indeed, condense the scene into a smaller area than a flat mirror would, and the dark color robs the scene of much of its color. It did not to me suggest the work of Claude Lorrain, or of any landscape painter. Rather, it made the image more like a *grisaille* painting, reduced to blacks and grays.

I also noticed two features of the scene as observed in a Claude Lorrain mirror. One is that the scene is "lively"—it slides around in your field of view very easily as you move your head or the mirror, and the image distorts slightly but rapidly as you do so. It's much the same effect one sees with a zograscope or perspective cabinet.[3] Viewers of images through such devices, which used large diameter, relatively short focal length lenses with a picture placed at the focal point, saw it as somewhat three-dimensional as they moved their heads from side to side. The Claude Lorrain mirror gives a similar impression of roundness and dimensionality that using a flat dark mirror would not.

The other effect is that the reduced image appears to be located near the focal point of the mirror, half a radius of curvature beyond the mirror surface. If you're myopic, as I am, you find that you can see a sharp image of a distant object without having to put on your corrective glasses.

I also had a theory regarding the plentiful mirrors sold to all those dilettantes. Just as today we have a relatively few high-end telescopes and binoculars with plastic optics, and many more of lesser quality and lower price, so must the Claude Lorrain mirrors have the same classes of quality. I suspected that most such mirrors would be of poor quality, badly ground and polished, with an indifferent shape that was closer to a flat rectangular slab with rounded-off sides than to a section of a true sphere. Since the item is not a piece of transparent optical glass, but a dark and opaque material, its shortcoming would be less immediately obvious.

Professor Jamie Day of the Physics Department at Transylvania University in Kentucky was kind enough to send me some data regarding the Claude Lorrain mirror they have in the University's collection, and I must admit that their mirror does have a good surface figure, without the gross mishapings that I had feared. Of course, the preserved specimens we have are probably drawn from the ranks of the best ones made, but it's something of a relief to find them of such good quality.

I'd like to thank those who helped me with this investigation, including Professor Day, Deborah Warner of the Smithsonian Institution (which also has a Claude Lorrain mirror), and artist Alex McKay, who has done much work with such mirrors and has a website devoted to them (with a live feed from a camera looking into one).[4] As usual, any errors in this piece are my fault and responsibility, and not due to the people who generously shared their time with me.

Notes

1. Henry Clay Trumbull writing about the use of the Claude Lorrain mirror in *Seeing and Being: Or, Perception and Character*, Chapter XIII, "The Softening Light of Reflection" (Philadelphia: John D. Wattles, 1889), 132–133. https://books.google.com/books?id=D8kV AAAAYAAJ&pg=PA131&dq=%22Claude+Lorraine+Glass%22&hl=en&sa=X&ei=c66JUa mqCarQ0wH7voH4Ag#v=onepage&q=%22Claude%20Lorraine%20Glass%22&f=false

2. See Stephen R. Wilk's Light Touch column, "With the Camera Lucida You, Too, Can Be an Artist," *Optics & Photonics News* 19, no. 7 (July 2008): 14–17; or Chapter 12 ("Even If You Can't Draw a Straight Line …") in *How the Ray Gun Got Its Zap!* by Stephen R. Wilk (New York: Oxford University Press, 2013).

3. Stephen R. Wilk, *Light Touch* column "Zograscopes," *Optics & Photonics News* 23, no. 7(July/ August 2012): 16–17; or Chapter 38 in this volume.

4. http://web2.uwindsor.ca/hrg/amckay/Claudemirror.com/Claudemirror.com/Entrance. html (accessed February 11, 2016).

40
How the Ray Gun Got Its Zap! Part II—Handheld Ray Guns

Chapter 36 in my earlier book *How the Ray Gun Got Its Zap!* explained the origins and history of the ray gun, but it didn't cover what might be called an essential part of its development.

To recap, weapons relying upon electromagnetic rays are as old as scientific investigations into those rays. Astronomer William Herschel is credited with the discovery of infrared rays in 1800. Johann Wilhelm Ritter discovered ultraviolet rays at the other end of the spectrum a year later. What is arguably the first ray gun made its appearance just eight years later in a book by American author Washington Irving. It was in his comic work *The History of New York from the Beginning of the World to the End of the Dutch Dynasty*, ascribed to Diedrich Knickerbocker.[1] (The book is usually called simply *Knickerbocker's History* by those wanting to avoid lengthy titles, but that misses the joke implicit in the long form.)

Chapter Five, Part One of this work contains an unexpected detour into speculation. In defending the original inhabitants of North America, it asks what would happen if Europeans had been subjected to a similar invasion, but from outer space:

> To return, then, to my supposition—let us suppose that the aerial visitants I have mentioned, possessed of vastly superior knowledge to ourselves—that is to say, possessed of superior knowledge in the art of extermination—riding on hippogriffs—defended with impenetrable armor—armed with concentrated sunbeams, and provided with vast engines, to hurl enormous moonstones; in short, let us suppose them, if our vanity will permit the supposition, as superior to us in knowledge, and consequently in power, as the Europeans were to the Indians when they first discovered them.

Irving's invaders came from the Moon, and his equipping them with "concentrated sunbeams" not only imagined a weapon vastly different from European guns, but actually made use of recent science. Two years previously, Francois Peyrard's book *Œuvres d'Archimède, traduites littéralement, avec un commentaire par F[rançois] Peyrard* included a discussion of Archimedes's mirror, the parabolic weapon he was said to have used to concentrate sunlight onto the invading Roman fleet at the Siege of Syracuse in 313 BCE, and people were attempting to duplicate the feat.

Almost 90 years later, an eerily similar story was told by Herbert George Wells in *The War of the Worlds*, where Wells also explicitly drew the parallels between Europeans being attacked by technologically superior invading aliens (from Mars this time), and the way the Europeans themselves had invaded the Americas and other less-developed portions of the world. Again, the invaders were given a ray weapon, a heat ray that the Europeans had no defense against. Wells very properly imagined the ray itself being invisible until it struck a target, but his illustrators, right from the original serialization in *Pearson's Magazine*,[2] were unable to resist showing the rays travelling from the Martian machines to their targets. Wells described the possible mechanism as an infrared source, directed in a collimated beam using a parabolic reflector. (The author has, in fact, used such an arrangement for measuring the infrared imaging properties of lenses, rather than for destruction.) Wells's description, incredibly, describes the way an infrared laser operates, and such infrared weapons are, at long last, starting to become realistic beam weapons on the world stage today.

After Wells, others invented and used a variety of beam weapons. Garrett P. Serviss, who edited an unauthorized publication of Wells's book in US newspapers, could not resist writing a sequel in which Earth people (and Americans in particular) struck back against the Martians with their own beam weapon, the "disintegrator" (to my knowledge, the first use of that term in this context), with *Edison's Conquest of Mars* (1898).[3] Edison's disintegrator worked by exciting the atoms of the object it was aimed at. The device could be adjusted in frequency, and, it was claimed, every substance had its own characteristic frequency of vibration. Exciting the object with this frequency caused increased oscillations at resonance, causing the item ultimately to fly apart. Exactly where this idea came from, I cannot say. Louis de Broglie wouldn't introduce the concept of matter waves for another quarter century. But Edison, at least in Serviss's book, was using them in 1898 to destroy a bird:

> We had gone upon the roof of Mr. Edison's laboratory and the inventor held the little instrument, with its attached mirror, in his hand. We looked about for some object on which to try its powers. On a bare limb of a tree not far away, for it was late in fall, sat a disconsolate crow.
>
> "Good," said Mr. Edison, "that will do." He touched a button at the side of the instrument and a soft, whirring noise was heard. "Feathers," said Mr. Edison, "have a vibration period of three hundred and eighty six million per second.
>
> He adjusted the index as he spoke. Then, through a sighting tube, he aimed at the bird.
>
> "Now watch," he said.
>
> Another soft whirr in the instrument, a momentary flash of light close around it, and behold, the crow had turned from black to white!
>
> "Its feathers are gone," said the inventor; "they have been dissipated into their constituent atoms. Now, we will finish the crow."
>
> Instantly there was another adjustment of the index, another outshooting of vibratory force, a rapid up and down motion of the index to include a certain range of vibrations, and the crow itself was gone—vanished in empty space!

For years afterward disintegrators were the ray guns of choice, originally said to operate in this vague manner, and later simply invoked, without any explanation. A succession of such ray machines appeared in works of fiction. George Chetwith Griffin-Jones's *The World Masters* (1903)[4] had a device called both a "disintegrator" and a "death ray" (possibly the first use of that lethal term). His 1911 novel *The Lord of Labour*[5] described a future battle fought with disintegrators and atomic missiles. Victor Rousseau's *The Messiah of the Cylinder* (1922)[6] has ray guns in the future. Percy

F. Westerman's *The War of the Wireless Waves* pitted British ZZ rays against German Ultra-K waves.[7] Alexei Tolstoy's *The Hyperboloid of Engineer Garin* features an engineer who invents a ray gun and tries to use it to become dictator.[8] There were many others in novels and in films.

The August 1928 issue of the pulp magazine *Amazing Stories* was historically significant. It featured Edward Elmer "Doc" Smith's first story (written in collaboration with Lee Hawkins Garby, who is rarely given credit for her contribution), "The Skylark of Space," the archetypal "space opera," which Smith and writer Edmond Hamilton made their specialty—a far-ranging genre that eventually lead to such items as the *Star Wars* films. The same issue featured the story "Armageddon 2419 A.D." by Phil Nowlan which led to the same end by a different route. The novel featured a future in which earthly adversaries fought wars with what are recognizably the same disintegrators, projected from flying machines.

All of these ray guns and disintegrators had one feature in common: they were big. They were more ray cannons than ray guns. Edison's handheld disintegrator that he used on the crow must be seen as a miniature prototype, because later in Serviss's novel they are using the same sort of ray cannons. All such items were placed in stands, or on sturdy mounts, and were often far too heavy to lift. Nobody was using portable handheld ray weapons. There were no ray rifles or ray pistols.

All of that was about to change, and Nowlan's story, along with its sequel, *The Airlords of Han*,[9] was to be the instrument of that change. The latter was, as I pointed out in my previous book, the source of the onomatopoeic sound "Zap!," which has become the hallmark of the ray gun. The model appears to have been high-voltage electricity, because discharges of high voltage do, indeed, make a "zap" sound. Early ray guns were also often depicted with the same sort of ribs characteristic of insulators, which reinforces the analogy with electricity.

The hero of Nowlan's stories was Anthony Rogers, a hero in the mold of many of those in such "scientific romances." They were often military officers and/or wealthy individuals: Edgar Rice Burroughs's John Carter of Mars, Edwin Lester Arnold's Lieutenant Gulliver Jones, and "Ralph Milne Farley's" (pen name of Roger Sherman Hoar) Myles Cabot of Venus. Jones, too, was a veteran of World War I, who falls into a state of suspended animation and awakes in the year 2419, where he finds that inhabitants of America are at war with the Han, a vaguely Asian people, armed with flying machines and disintegrator cannons. The stories were successful, and Nowlan sought to continue his series not by writing further stories, but by serializing them in the form of the newly popular newspaper comic strips. He took his idea to distributor John F. Dille of the National Newspaper Service syndicate, who got Dick Calkins to draw a daily strip and Russell Keaton to draw the color Sunday pages.

They reused Nowlan's stories to launch the strip, but they made one change. It's reported to have been the idea of Dille to use instead of Rogers's first name "Anthony," the nickname "Buck," possibly inspired by cowboy star Buck Jones. The change possible contributed to the strip's success, because instead of the patrician Anthony Rogers (like all those other heroes), the hero was now the approachable everyman "Buck" Rogers.

At first the fantastic elements were less fantastic—the stories were set on the future Earth, with flying machines (and flying people), disintegrator cannons, and the like. The people Buck encounters use "rocket pistols" that fired explosive bullets. But gradually, the plots got more fantastic. Buck Rogers started using space ships, and adventures began to be set off Earth, on other planets in the solar system. And instead of using pistols with explosive bullets or disintegrator cannons, Buck and company acquired handheld ray funs. First, there were rather bulky disintegrator rifles and handheld disintegrator pistols in the hands of the Mongols, but eventually Buck's group acquired them, too. An icon was born.

People probably weren't aware of its significance at first—there was so much new and futuristic in Buck Rogers that this detail was lost in the shuffle. Comic strips were the pop culture medium of the day, and they could present the fantastic much more convincingly that the primitive motion pictures of the time. Buck Rogers caught on in popularity, the first science fiction comic strip, and soon branched out into a radio show in 1932, then into comic books (that outgrowth of the newspaper strips) in 1933. A short film of Buck Rogers was made in 1933 for the Chicago World's Fair, starring Dille's son. This led, eventually, to a movie serial.

Buck's iconic ray gun achieved a popularity of its own. First there appeared paper models, associated with product premiums and newspapers, but within a year the Daisy company, makers of the popular "BB" guns that shot tiny metal pellets, started making various models of Buck Rogers ray guns. At first they sold the XZ-31 Rocket Pistol, just as they used at first in the strip, with its explosive bullets. But they soon produced the very similar-looking XZ-38 Disintegrator. It reportedly made a "Zap!" sound upon being fired.

The ray gun, as a concept, was now firmly established. It was a completely new concept—the handheld futuristic weapon. In addition to the wish-fulfillment fantasy it presented, the ray gun served other functions. It provided a "science fiction" equivalent of the detective's or cowboy's gun. It identified the genre immediately and was appropriate to it. Just as you needed a tractor beam to have one space ship pull another (why would you use something as mundane as a rope or cable? This is science fiction!), so it was appropriate that spacemen and spacewomen would use a tool appropriate to their trade and milieu, the ray gun.

Another reason, I believe, became increasingly important as the 1930s continued toward war and into World War II in the 1940s. Ray guns allowed the heroes and villains of popular escapist fiction to shoot at each other without wounds or blood, thus avoiding upsetting people on the home front by not reminding them about the war. It's similar to the way that, at the same time, dissolution by sunlight became a preferred method of destroying vampires (see Chapter 31).

The success of Buck Rogers inevitably inspired imitators. "Brick" Bradford was introduced in the comic strips in 1933, just when Buck Rogers became popular. He, too, started as an earthbound aviator but soon was having interplanetary adventures and using ray guns.

"Flash" Gordon was started in 1934 by Alex Raymond in direct response to the popularity of Buch Rogers. Originally King Features syndicate, who ran "Flash," wanted to adapt Edgar Rice Burroughs's "John Carter of Mars," but they couldn't come to term on the rights, so Raymond invented his own story, cribbing elements from John Carter, from Buck Rogers, and from the Philip Wylie and Edwin Balmer novel *When Worlds Collide*. This explains why Flash Gordon anachronistically fights with swords (like John Carter) and ray guns (like Buck Rogers). John Carter dates from an era when military officers used firearms but still carried ceremonial swords. Burroughs's Martians also used swords and rifles. In 1912, when his stories first appeared, the handheld ray gun hadn't been thought of yet. Burroughs eventually introduced ray guns into his series, but only after others had popularized the idea.

Actually, there had been one earlier incarnation of the handheld ray gun. It appears in Abraham Merritt's 1919 story "Conquest of the Moon Pool," a sequel to "The Moon Pool" that had appeared a year earlier.[10] The ruins of the lost civilization from the first story lead to the discovery of a still-inhabited Lost City in the second. The two stories were combined into a novel published in 1919. The evil queen Yolara uses a handheld disintegrator called a *keth* to eliminate an annoying antagonist:

> "As for you—you have lived long enough, Songar! Pray to the Silent Ones, Songar, and pass into the nothingness—you!"
>
> She dipped her hand into her bosom and drew forth something that resembled a small cone of tarnished silver. She levelled it, a covering clicked from its base, and out of it darted a slender ray of intense green light.
>
> It struck the old dwarf squarely over the heart, and spread swift as light itself, covering him with a gleaming, pale film. She clenched her hand upon the cone, and the ray disappeared. She thrust the cone back into her breast and leaned forward expectantly; so Lugar and so the other dwarfs.... For the moment the white beard stood rigid; then the robe that had covered him seemed to melt away, revealing all the knotted, monstrous body. And in that body a vibration began, increasing in incredible rapidity. It wavered before us like a reflection in a still pond stirred by a sudden wind. It grew and grew—to a rhythm whose rapidity was intolerable to watch and still chained the eyes.
>
> The figure grew indistinct, misty ... The glowing shadow vanished, the sparkling atoms were still for a moment—and shot away, joining those dancing others.
>
> Where the gnomelike form had been but a few seconds before—there was nothing!

Merritt was, at the time, an immensely popular author, but no one caught up this image of a handheld disintegrator. There was no hugely popular figure to be imitated, as with Buck Rogers, and so this first appearance of a handheld ray gun died aborning.

The idea and the image of the handheld pistol-like ray fun (and its logical extension, the rifle-like ray gun) was now firmly established as an instantly recognized element of science fantasy and science fiction. Since the essence of good science fiction is extrapolation and examination of the consequences of new ideas, the proper thing to do was to treat this new concept the same way. If ray guns were a reality, how would they affect society?

Robert Heinlein answered that question with his story *Beyond This Horizon*.[11] It was then expanded into a novel and published in 1948. It's a look at look at a future utopia where most of humankind's ills have been conquered. Dueling has been reintroduced into society, with elaborate norms and rituals, and the weapon used is the "needlebeam," a form of ray gun. Purportedly the dueling society was introduced in response to a request from *Astounding* editor John W. Campbell to illustrate his belief that "an armed society is a polite society" (although that was arguably a Heinlein dictum as well), but the choice of weapon was Heinlein's, and it made a good use of the new trope. The needlebeams are small and discreet and make minimal mess. At one point the protagonist, Hamilton Felix, meets with a government official, Mordan, who avoids the use of such weapons, although he wears one at his side.

Hamilton chewed his lip. "I say … you'll pardon me … but isn't it indiscreet for a man who does no fighting to appear in public armed?"

Mordan smiled. "You misconstrue. Watch." He indicated the far wall. It was partly covered with a geometrical pattern, consisting of small circles, all the same size and set close together. Each circle had a small dot exactly in the center.

Mordan drew his weapon with easy swiftness, coming *up*, not down, on his target. His gun seemed simply to check itself at the top of its swing, before he returned it to its holster.

A light puff of smoke drifted up the face of the wall. There were three new circles, arranged in tangent trefoil. In the center of each was a small dot.

Hamilton said nothing. "Well?" inquired Mordan.

"I was thinking," Hamilton answered slowly, "that it is well for me that I was polite to you yesterday evening."

Around the same time, author Alfred E. Van Vogt wrote a series of stories later collected as *The Weapon Shops of Isher*,[12] which argued strongly in favor of the right to bear arms. "The Right to Buy Weapons Is the Right To Be Free" is the motto of the titular shops. The weapons are, of course, ray guns, but they are configured so that they can only be used in self-defence. The weapon shop motto has been quoted favorably by the National Rifle Association in the United States.

The idea of a ray gun wasn't purely a literary invention. From 1900 to 1939 more than two dozen inventors claimed to have developed ray guns or death rays, which they tried to market to the military. Some of these were taken seriously enough to be tested, and many of them ended up on the covers of magazines like *Popular Science*. This gave the concept of such superscientific devices a veneer of respectability, although none of them ever panned out as a successful addition to the arsenal of the US military.

Not that the writers would let mere inventors get the jump on them. As writer and critic James Blish recalled of the Golden Age of Science Fiction:[13]

We had rays that would kill you by coagulating your proteins, as though you were a hard-boiled egg. We had rays that would carry a deadly electrical shock, of course—that was beginner's stuff—and poison rays which would turn your blood into furniture polish. We had several different types of disintegrator, which either made you vanish completely or turned you into fine dust or pocket-flug; … We had rays which would drive you insane; rays which would throw you into convulsions; rays which would paralyze you; rays which would melt you down like a tallow candle. Harl Vincent invented one which covered you with hundreds of buzzing, spinning little black discs, which wore you rapidly down to nothing but a curl of greasy smoke…. And of course we had heat rays, from Wells onward; and Ray Cummings had a cold ray, too.

For most writers, however, ray guns soon became the normal accepted prop of futuristic fiction, acting in the same position as a six-gun in Western fiction, and used with little concern about its mechanism. So when Catherine L. Moore introduced her cowboy-like hero Northwest Smith in 1933's *Shambleau*, all she has to say is that he carries a heat gun, and that's all you need to know.[14] But whether it was a heat gun, a ray gun, a blaster, a needlebeam, a tickler, or, much later, a phaser, it was the all-purpose science fiction sidearm.

As the years went by, others continued to look at the ramifications. Robert Sheckley ruminated on the silence of the ray gun in the 1958 story "The Gun Without a Bang,"[15] and felt that animals wouldn't find it as effective as a gun using normal bullets because of the lack of a satisfactory "bang." The makers of the *Rifts* role-playing game imagined that their ray weapons could be fitted with sound generators to produce an "intimidation effect."

Ray guns came to be looked at satirically. In the cartoon *Duck Dodger in the 24 ½ Century* (1953), directed by the science fiction–savvy Chuck Jones, the titular Duck Dodgers (Warner Brothers cartoon staple Daffy Duck doing a Buck Rogers parody) is zapped by Marvin the Martian with an "A-1 Disintegrating Pistol" (which bears a suspicious resemblance to Buck Rogers's disintegrator). But the disintegrated Duck Dodgers is restored when his sidekick uses the obvious antidote to the disintegrating pistol—the "Acme Integrating Pistol."

A decade later science fiction writer Philip K. Dick was approached by Pyramid Books, who wanted him to do an up-to-date take on the ray gun, considering the scientific advances that made it look more like a reality. But Dick's book was a broadly satirical novel, *The Zap Gun*,[16] which posited that the apparent ray gun arms race was actually a sham. Instead of newer and better ray guns, the new technology is directed toward consumer products, ray gun "swords" beaten into economic "plowshares."

Inevitably, the engineering types that contribute to the science fiction world took a critical look at these devices and observed their limitations. In *The Sands of Mars* by Arthur C. Clarke (1952),[17] scientists on Mars are discussing what is really an ultra-bright flash system, rather than a ray gun. But their point is still germane:

> "Let's have a look at that thing," said Hilton.
> Gibson handed over the flash-gun and explained its operation.
> "It's built around a super-capacity condenser. There's enough there for about a hundred flashes on one charge, and it's practically full."
> "A hundred of the high-powered flashes?"
> "Yes; it'll do a couple thousand of the normal ones."
> "Then there's enough electrical energy there to make a good bomb in that condenser. I hope it doesn't spring a leak."

And that, of course, was an unspoken and generally unrecognized problem with ray guns—the huge electrical energy needed to produce and project energy sufficient to do damage and stored in the gun itself was also potentially dangerous to the user, should the charge all be released at once. The TV series *Star Trek* recognized this—in the series a "phaser" could be made to "overload," turning it into a miniature bomb.

And then, of course, the laser came along and ruined everything.

Who would have predicted that an outgrowth of microwave resonator work, placing an optical gain material inside a resonant cavity, would succeed in producing a real-life analog to Wells's Martian heat ray? All those ideas of collimated beams from point sources reflecting from parabolic reflectors, like ultra-searchlights, were replaced with Gaussian beams that delivered energy much more effectively. The public, as soon as it saw the laser, was ready to take it for the real thing, the ray gun made real. Hadn't they shown that the ruby laser could drill a hole in a diamond? Couldn't Arthur Schawlow use it to burst a blue balloon inside a clear one without harming the outer balloon?

Alas, there were a great many bugs to work out first. The gain media were woefully inefficient (necessarily, in order to generate a population inversion), and as a result much energy was lost to heat. Powerful lasers required massive amounts of cooling, as well as massive electrical supplies. The military, also eager for recoil-less weapons that didn't have to "lead" a target, was also disappointed but its finickiness and impracticalities. Some concluded that it would be more efficient to simply drop the huge laser devices on the enemy than to try and shoot the beams at them.

One way around the problem of large electrical supplies, heat, and the limitations of storage batteries was to use pyrotechnic sources to pump the laser gain material. In the 1980s, the Soviet Union really did design and build handheld laser "pistols" using pyrotechnic cartridges to produce short, intense laser beams. The pyrotechnic flash lamp was operational by 1984 and was intended to blind optical sensors. It couldn't really do any other damage. It was never used, although several models were constructed.

But ray guns continued to be popular in pop culture, and real-life lasers got incrementally better. Advances in solid-state lasers and beam-combining technology have made military lasers capable of doing real damage a practical reality. For the most part, however, these remain tethered to large power supplies—laser cannons, rather than laser pistols. There are reports that the Chinese have produced a battery-powered laser "rifle" that can set fires from a distance. Home laser hobbyists have bundled together large numbers of lithium ion batteries with laser cavities to produce laser pistols and rifles that can do the same thing, and they have posted their triumphs on the Internet and videos of them in operation on YouTube.

But the reality is that these are still exotic curiosities. They can do unexpected damage to the unprepared, but the fact that it requires huge amounts of optical energy to do serious damage to water-filled animal tissue or to melt metal means that these portable laser devices are still very, very far from the capabilities of the fictional ray guns.

Notes

1. Diedrich Knickerbocker (Washington Irving), *The History of New York from the Beginning of the World to the End of the Dutch Dynasty* (New York: Inskeep and Bradford, 1809).
2. Herbert George Wells, *The War of the Worlds*, serialized in eight parts in *Pearson's Magazine* (April 1897–December 1897) and in *The Cosmopolitan* in July and August 1897; published as a single volume by William Heinemann (London, 1898).

3. Garrett Serviss, *Edison's Conquest of Mars*, originally published serially in *The New York Journal* (January 12, 1898 to February 10, 1898), later in book form by Carcosa House (1947). Scans of first publication online here—http://durendal.org/ecom/index.html

4. George Chetwith Griffin-Jones (as George Griffin), *The World Masters* (London: John Long, 1903).

5. George Chetwith Griffin-Jones (as George Griffin), *The Lord of Labour* (London: F.V. White and Co., 1911).

6. Victor Rousseau, *The Messiah of the Cylinder* Serialized in *Everybody's Magazine* June-September 1917 (published in a single volume by Chicago: A.C. McClurg and Co., 1917).

7. Percy F. Westerman, *The War of the Wireless Waves* (New York: Oxford University Press, 1923).

8. Alexei Nikoleyevich Tolstoy, *The Hyperboloid of Engineer Garin* (London: Methuen, 1927). Tolstoy was a bit off in his solid geometry. He surely meant a *paraboloid*, not a *hyperboloid*, since he visualized a mirror of that shape collimating the beam, as with Wells's Martian heat gun. A hyperboloid won't do this.

9. *Amazing Stories* (Vol. 3, No. 12, March 1929, 1106–1136). Actually, I have learned while proofing this book that the "Zap!" actually first appears in the comic strip adaptation, *Buck Rogers 2429 A.D.* dated October 21, 1929. I thank Steve Carper, science fiction historian extraordinaire, for this tidbit.

10. Abraham Merritt, "The Moon Pool," *All-Story Weekly* (June 22, 1918). "Conquest of the Moon Pool" in *All-Story Weekly* in six parts (February 15, 1919 through March 22, 1919). Collected and revised as a single volume (New York: G.P. Putnam's Sons, 1919).

11. Robert A. Heinlein, *Beyond This Horizon*, serialized in the premier science fiction magazine *Astounding*, in April and May 1942, as by "Anson McDonald." Published as a single volume by Fantasy Press in 1948. Some later paperback editions prominently feature the needlebeam weapons on the cover. The first instalment in *Astounding* didn't, however. Campbell had a long-standing dislike of ray guns on magazine covers, evidently thinking that it cheapened the medium. Heinlein's novel was long eclipsed by his other works but was awarded a "Retro-Hugo" award in 2018 and has of late been critically reconsidered.

12. Alfred E. Van Vogt, *The Weapon Shops of Isher*. Portions appeared as short stories in 1941–1949. Combined and re-edited into a novel in 1951 by Greenberg.

13. James Blish, *The Issue at Hand* (Chicago IL: Advent, 1964), 45.

14. Catherine L. Moore, "Shambleau" originally appeared in *Weird Tales*, November 1933. Reprinted frequently thereafter in anthologies.

15. Robert Sheckley, "The Gun Without a Bang," originally published in *Galaxy Science Fiction* (June 1958) and frequently anthologized afterward.

16. Philip K. Dick, *The Zap Gun*, published originally in two parts in *Worlds of Tomorrow*, November 1965 and January 1966. Combined into a single book by Pyramid Books in 1967.

17. Arthur C. Clarke, *The Sands of Mars* (New York: Gnome Press, 1952)

Afterword

That's the latest harvest of my pieces on unusual optics, including some that have not appeared before in print. As I stated in my first book, I am always looking for instances where some optical effect seems to go against the general understanding, uses unusual materials, or where the topic lies at the intersection of optics and popular culture. And I always look for the lessons that might be learned from these mini-investigations. I hope to make people look at things in a new light and question phenomena and assumptions they have taken for granted.

In the afterword to *How the Ray Gun Got Its Zap!* I asked what sort of a color "brown" was. It's clearly not on my CIE chromaticity diagram. But that's supposed to cover all possible colors, so where is brown? I address that in the present book (Chapter 16), and in the process we learn a little more about how colors were organized and defined. Similarly, asking about the values of gray on the standard image charts, we look into how those "steps of equal gradation" were defined. There are plenty of other odd questions to be asked. What, for instance, causes those shimmering bands of light that appear to emanate from the shadow of my head when I look into a turbid body of water?

My query about why ultraviolet light kills vampires might appear frivolous, but as I ask in the introduction, how did that notion develop in the collective consciousness of a public that doesn't really think or care much about spectroscopy and its implications? The answer is that it was pushed along by government warnings about the dangers of overexposure to direct sunlight and the promotion of the idea of solar protection factors (SPFs). Knowing that, you're equipped to study how other scientific and engineering concepts might enter public discourse, or even how to try to use them yourself in education.

I give lectures at schools and organizations, and every year I run a "hands-on science" session for grade-school kids at a convention. I've had two such sessions make "x-ray specs" (Chapter 35). It's not only easy to do, but it lets you ask how the apparent "x-ray" image is formed, either using feathers or phase structures, and it lets you talk about both. Even if the kids don't fully understand how the phenomenon works, they've at least seen it in action, and so can see a bit more of the world of physics. Similarly, when I have them build spectrographs out of old CDs or DVDs and paper towel rolls, it's not so much important that they understand why the disc splits up light (I limit my technical explanations to the length of a TV commercial) as that it actually does do so. Again, they understand that the phenomenon of diffraction gratings exists, and they can come back to understanding it later. More important, they now know about spectroscopy, and that light can be broken up into constituent colors, and that you can learn things from what those colors look like and how strong they are relative to each other.

Even if optics is your profession, there are undoubtedly parts of it that have never come under your particular study, job, or interests, and it's worthwhile looking at these. Why is a candle flame yellow? The explanation I'd been given—it's blackbody radiation—isn't by any means the whole story, and it misses the important fact that soot in micron-sized particles doesn't act as a blackbody. Why is a fogbow white and a glory colored, if they're the result of drops of the same size? Why would anyone try to correct vision in both eyes with a single lens? How could the zograscope produce what its proponents claimed were three-dimensional images with a single image and a single lens? What did anyone get out of a "Claude Lorraine" mirror, if you couldn't record the image?

The questions are always worth asking, and the way you get to the answers, as much as the answers themselves, can help you when inquiring about your own problems.

Index

For the benefit of digital users, indexed terms that span two pages (e.g., 52–53) may, on occasion, appear on only one of those pages.

Boxes are indicated by b following the page number